The
Black Hills

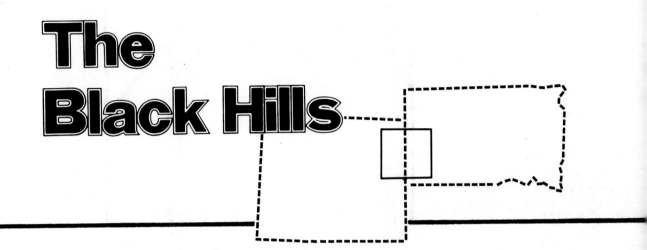

K/H GEOLOGY FIELD GUIDE SERIES

FIELD GUIDE

The Black Hills

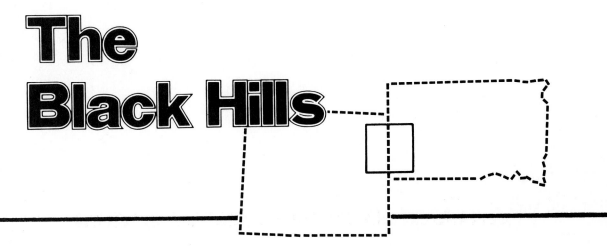

Rodney M. Feldmann
Richard A. Heimlich

Kent State University
Kent, Ohio

KENDALL/HUNT PUBLISHING COMPANY
Dubuque, Iowa, USA • Toronto, Ontario, Canada

K/H Geology Field Guide Series

Consulting Editor
John W. Harbaugh
Stanford University

Printed in the United States of America

C 402193 01

CONTENTS

Foreword vii

Preface ix

Acknowledgments xi

1. Introduction 1

2. Geological Setting 7

3. Locality Descriptions 43
   Stop 1. Lookout Peak and the Early Mesozoic Rock Sequence, Spearfish 43
   Stop 2. Bridal Veil Falls, Spearfish Canyon 46
   Stop 3. Bear Butte 51
   Stop 4. Sly Hill Overlook, Sturgis 56
   Stop 5. Late Paleozoic Rocks and the Boulder Park Syncline 59
   Stop 6. Sheep Mountain Stock 65
   Stop 7. Whitewood Creek Gold Desposit 65
   Stop 8. The Deadwood Formation 70
   Stop 9. Winnipeg, Whitewood, Englewood, and Pahasapa Formations 75
   Stop 10. Cutting Stock 80
   Stop 11. Terry Peak 83
   Stop 12. Precambrian Rock Units in the Lead Area 86
   Stop 13. The Homestake Gold Deposit 94
   Stop 14. Strawberry Ridge Iron Deposit 100
   Stop 15. Strawberry Ridge Schist 104
   Stop 16. Cenozoic Volcanic Rocks 105
   Stop 17. Nemo Iron Deposit 109
   Stop 18. Little Elk Creek Canyon—White Gate 112
   Stop 19. Dinosaur Overlook 117

Stop 20.   Peerless (Rushmore) Pegmatite   120
Stop 21.   Etta Pegmatite   124
Stop 22.   Mount Rushmore National Memorial   129
Stop 23.   Iron Mountain Pegmatite   139
Stop 24.   The Needles   139
Stop 25.   Jewel Cave   144
Stop 26.   Custer Schist   149
Stop 27.   Bull Moose Pegmatites   153
Stop 28.   Buffalo Gap   157
Stop 29.   Hot Spring Mammoth Site   161
Stop 30.   Cascade Springs   164

Glossary   171

References   183

Index   185

FOREWORD

This book is one of the K/H *Geology Field Guide Series*. The objective of this series is to provide an authoritatively written layman's guide to important geologic features in each region treated. Stress is placed on observations in the field. Each guide provides an overview of the region to which it pertains, and outlines a series of self-guiding field trips which will allow users to make their own first hand observations of features that typify the area.

The series is directed toward diverse groups of users. It should find use in formal classes in geology, both at the college and university level, and in high schools, in which field trips form an essential part of introductory or advanced courses. Furthermore, the books in the series should be useful to professional geologists and other scientists who desire an introduction to the geologic features of particular regions. Finally, they should find use among individuals who are not necessarily trained in science, but who do have an active interest in natural history and who enjoy travel.

Authors of these books all have intimate acquaintance with their respective regions and extensive teaching experience which has stressed field trip observations. Consequently, each book represents a distillation of teaching experiences that have involved many students and numerous field trips.

John W. Harbaugh
Consulting Editor

PREFACE

Geology is a "hands on" subject. One experiences very little of the excitement of earth study simply by reading books, watching natural history programs on television, or attending lecture classes. To fully appreciate the features developed in the earth's crust, it is necessary to examine them first hand, puzzle over their identity, and attempt to interpret their origin. Possessing a basic interest and motivation, as well as the ability to apply relatively few geologic concepts, one can readily determine the significance, in various rocks, of diverse fossils, many representing organisms that have long since become extinct. Similarly, one can ponder and begin to understand the occurrence of beautiful minerals concentrated in solution cavities or as huge crystals in solid rock.

Many books have been written for the purpose of introducing geological principles and of providing information necessary for identification of rocks, minerals, and fossils. Our attempt has been to provide an overview of the geology of the Black Hills, to guide interested amateurs to points of specific geologic interest, and to describe briefly and illustrate the geological features at each locality. Our ultimate aim is to kindle an interest in the geology of the Black Hills and to provide a basis for appreciating the events in earth history that have shaped the area.

The Black Hills are ideally suited for this type of investigation because they include a vast array of geological features in a relatively small area. A distinct advantage is that most of these features are on public land. Most visitors to the Black Hills devote the majority of their time to outdoor activity. It is our hope that such activity will include observation of the diverse and spectacular geological features which abound here and that this guide will facilitate an understanding of the geology of the area.

ACKNOWLEDGMENTS

Among those who read portions of the preliminary draft and made valuable suggestions are Drs. Jack A. Redden, Phil R. Bjork, and John P. Gries (South Dakota School of Mines and Technology), Mr. Gordon A. Nelson (Chief Geologist, Homestake Mining Company), and Drs. Donald F. Palmer and Peter S. Dahl (Kent State University). Each made valuable suggestions significantly improving the work but none should be held responsible for errors and shortcomings. Mr. Albert J. Hendricks (U.S. Park Service) made photographs of Jewel Cave available for reproduction herein. Photos of Mount Rushmore Memorial were also provided by the U.S. Park Service. Additionally, Gordon Nelson spent time in the field pointing out specific localities of interest and he provided copies of several unpublished maps for use in the guide. Finally, we are grateful to Mr. Steve A. Heimlich who served as a cheerful and helpful field assistant.

Chapter 1

INTRODUCTION

Many areas of the United States are so well known for their scenic beauty that mention of them conjures up positive memories of past visits or hopes for future adventures. One such region is the Black Hills of South Dakota and Wyoming (fig. 1.1). Most of the areas that are popular vacation landmarks derive much of their scenic beauty from the local geology. Examination of brochures describing almost any state or federal park cannot help but impress the reader with the importance of geology in development of an area of special interest. The Black Hills is perhaps unique in that it is a relatively small area, about 120 miles long and 60 miles wide, in which geologic features are incredibly varied and readily available for examination.

The varied nature of the geology is, in one way, demonstrated by the fact that within the Black Hills are located a national park, a national monument, a national memorial, and three state parks, each devoted to the preservation of quite different types of geological features. Another indication of the diversity of the area lies in the observation that universities throughout the United States have established summer programs in the Black Hills to train geology students in field methods. Again, the variety of geologic features in the area makes this an ideal training ground. Finally, some of the charm of the Black Hills certainly lies in the distinct contrast between this region and the area surrounding it. The Black Hills have been described by many as a miniature Rocky Mountain system placed in the middle of the Great Plains. Indeed, when one travels westward across the northern tier of states, the Black Hills are the first mountain range encountered west of the Appalachians (fig. 1.2). Having crossed the Black Hills one again descends onto the Great Plains and must travel much farther west before mountains (the Bighorns) are again encountered.

Many Black Hills visitors have some degree of interest in geology, whether it is fascination with unusual rocks, minerals, and fossils, or whether it is confined to an aesthetic appreciation for the beauty of steep-walled canyons, waterfalls, caves, and mountain terrain. The Black Hills are ideally suited for these interests because the area is traversed by an excellent highway system (fig. 1.3) which brings many significant geologic features within easy reach of the traveler.

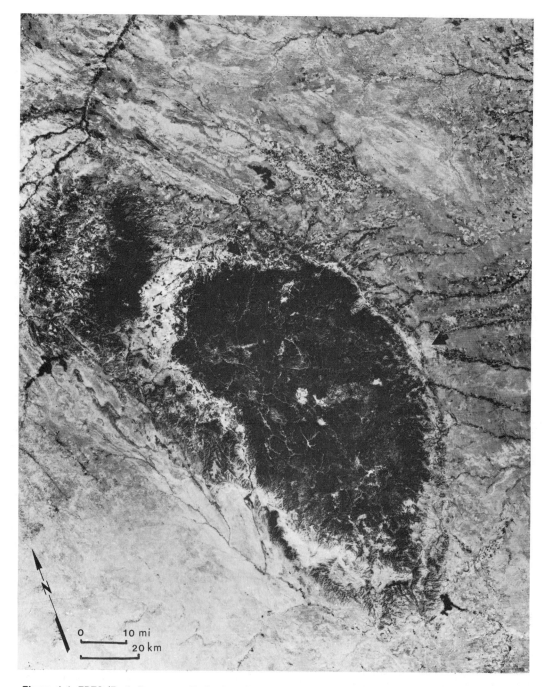

Figure 1.1. ERTS (Earth Resources Technology Satellite) photo of the Black Hills region. Arrows indicate the positions of Spearfish, South Dakota (near center of photo) and Rapid City, South Dakota. This satellite photo shows the structural pattern of the Black Hills, the major drainages, and some cities and highways.

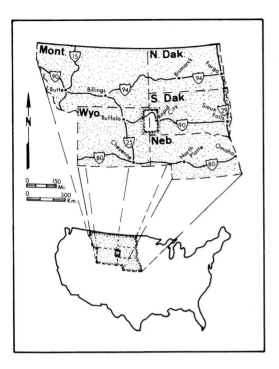

Figure 1.2. Generalized location of the Black Hills.

The purpose of this work is to introduce the interested amateur to some of the geologic materials comprising the Black Hills and to the processes that have formed the Hills. Hopefully, enough information is provided as well to permit a better understanding of the surficial processes that have shaped the landscape. The features of interest are illustrated and described in detail at numerous localities (fig. 1.3 and Table 1.1) that are readily accessible for examination, most without prior permission. Sites of particular geologic interest have been selected along nearly all the major highways crossing the Black Hills so that, no matter where one is in the region, described points of interest lie within only a few minutes travel. Needless to say, many equally interesting areas have not been described for the sake of brevity. The descrip-

tive material in Chapter 2 and the appendix will be of considerable assistance in interpreting the features not described at specific locations. The book should be of interest to itinerant geologists interested in examining more specific features of geologic significance.

Localities of specific interest are plotted on figure 1.3 and described in Chapter 3. Each of the stops is intended to be independent of the others so that there is no "proper" sequence in which they must be visited. Rather, the traveler can select as many points of interest as are practical and be assured that each will provide a unique insight into earth materials and processes. Various aspects of the geology detailed in the stop descriptions are outlined in Table 1.1. By selecting aspects of particular interest on the vertical axis, one can identify the specific stops where these features are displayed.

As is the case with any book of this sort, material has been extracted from many published sources and has been synthesized by the careful reading and criticism of many people. Primary geological references describing details of the geology of the Black Hills are surprisingly few. Among the more important are the monograph on the geology of the central Black Hills by Darton and Paige (1925), still the most comprehensive treatment of the area, and the U.S. Geological Survey (1975) compilation of the geology and resources of South Dakota which contains significant Black Hills geological summary articles by Norton, Redden, Gries, and others. In addition, a series of U.S. Geological Survey Professional Papers concerning the pegmatite deposits in the southern Black Hills provides detailed information about these unique features.

TABLE 1.1. Features of geologic interest plotted against locality numbers. The localities are described in detail in Chapter 3.

Feature at Locality

Locality Number	1	2	3	4	5	6	7	8	9	10	11	12	13	14	15	16	17	18	19	20	21	22	23	24	25	26	27	28	29	30
Pleistocene																												X	X	
Miocene																X														
Eocene		X	X			X				X	X			X				X									X			
Cretaceous			X	X														X									X			
Jurassic		X	X	X														X										X		
Triassic		X																X									X	X	X	
Permian					X													X	X									X		
Pennsylvanian					X													X												
Mississippian			X	X				X										X							X					
Devonian				X				X										X								X				
Silurian																														
Ordovician									X									X												
Cambrian		X						X				X	X					X												
Precambrian										X	X	X	X	X	X	X		X		X	X	X	X	X	X	X	X			

Chart of Paleontology and Special Features (columns unlabeled on this page; X marks indicate occurrence).

Category	Feature	X marks (left → right)
PALEONTOLOGY	Pleistocene	· · · · · · · · · · · · X
	Cretaceous	· X
	Jurassic	X
	Mississippian	· · X
	Devonian	· · · X
	Ordovician	· · · · X
	Cambrian	X · · · · X
SPECIAL FEATURES	Pegmatites	· · · X · · X X · · · X
	Iron Deposits	· · · · X · X · X
	Gold Deposits	· · X · · X
	Mining Activity	· · · X · X · X X
	Flooding	X X · · · · · X X
	Landsliding	X
	Ground Water	X X · · X
	Igneous Rocks	X X X X X X · X · · X
	Metamorphic Rocks	· · X X X X · X · X X
	Sedimentary Rocks	X X X X X X · X · X X X X
	Structures	· · X X · X · X · X X X

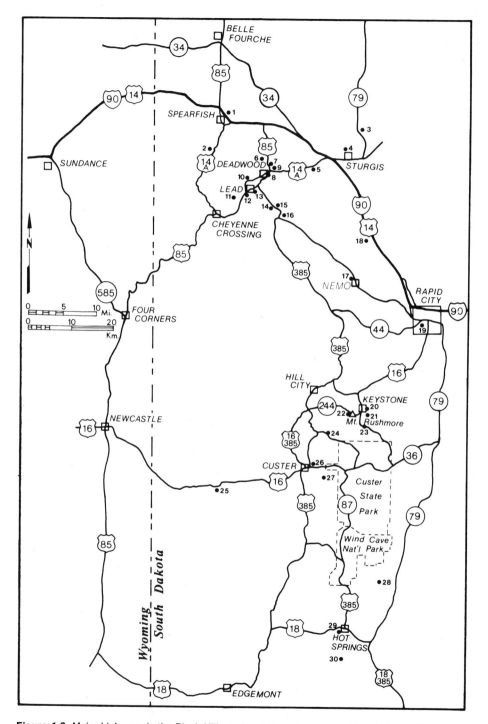

Figure 1.3. Major highways in the Black Hills region and localities described in the guide.

Chapter 2

GEOLOGICAL SETTING

MAJOR ROCK SYSTEMS

As one approaches the Black Hills from any direction the rocks exposed at the surface are of Mesozoic age, primarily Late Cretaceous (see geologic time scale, inside back cover). However, the Black Hills themselves are composed of considerably older rocks. These older rocks have been exposed at the surface by the processes of uplift and folding which formed the Black Hills; they have been exposed by erosion subsequent to that folding. Although the rocks which underlie the Black Hills are discussed more fully in the descriptions of individual stops, it is appropriate to summarize the general nature and distribution of these rocks here. They can be discussed best is "proper" geologic order, that is, from oldest to youngest.

The core of the Black Hills is exposed in an ovoid region extending from Lead to south of Custer (fig. 2.1). The rocks in this area are Precambrian in age and consist of two distinctly different groups. The older of the two is a sequence of metamorphic rocks which includes phyllite and schist along with smaller volumes of quartzite and amphibolite. These rocks, which have undergone extensive deformation, form the major part of the Precambrian core except in the south-central region where they have been intruded by younger Precambrian igneous material. The igneous rock of Precambrian age is referred to as the Harney Peak Granite. This granitic mass, and associated granitic pegmatite bodies, is visible in the areas of Mount Rushmore, Hill City, The Needles, and Custer. That the granitic material is younger than the phyllite and schist of the core can be demonstrated by observing that individual igneous bodies cut across, and were therefore injected into, the metamorphic sequence.

In turn, these rocks are all considerably older than the remainder of the rocks comprising the Black Hills. Examination of the contact of the Precambrian rocks with the overlying Paleozoic rocks indicates that, whereas the Paleozoic rocks are generally weakly folded, the Precambrian rocks have been intensely folded and faulted suggesting a long geologic history between the time that these rocks were originally deposited and the time of deposition of the Paleozoic sequence. This interface between the Precambrian rocks of the core and the Paleozoic rocks is referred to as a nonconformity. It represents a great interval of time, perhaps as much as one billion years, during which the Black Hills region was

7

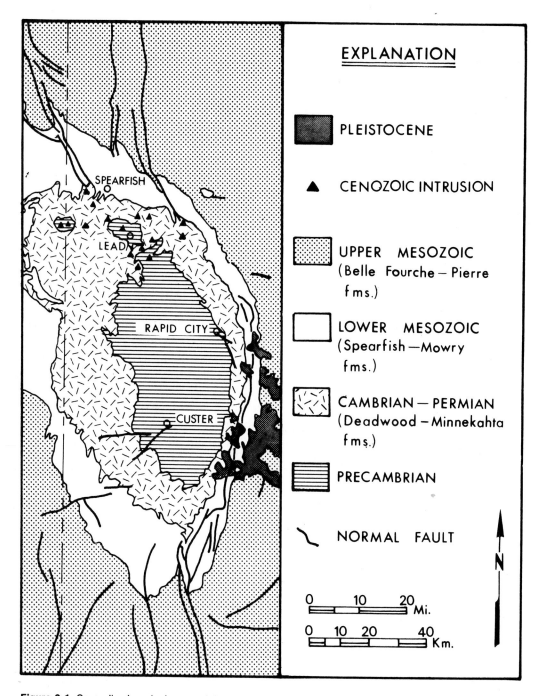

Figure 2.1. Generalized geologic map of the Black Hills showing distribution of rocks of various ages and positions of faults.

subjected to uplift and erosion. Mountains that may very well have existed in this region prior to that time were stripped down to a relatively low-relief surface on which Paleozoic sediments were deposited.

Rocks of Paleozoic age can be observed surrounding the Precambrian core. Most of these rocks were deposited under shallow marine conditions, a fact not necessarily implying that the process of formation of these rocks was continuous. In fact, it is perhaps more convenient to consider the Paleozoic sequence as being divided into three separate major units each of which is composed of different rock types and each of which formed under different depositional conditions. The earliest of these three includes the Cambrian and Ordovician rocks which are predominantly clastic silicate sedimentary rocks. This sequence marks the beginning of the Paleozoic history and is typical of much of the rock record throughout the upper midcontinent region. The rocks are relatively coarse-grained sandstones at the base of the sequence, overlain by finer-grained siltstone and shale, in turn overlain by a rather thin sequence of dominantly silty dolostone.

This sequence is separated from the overlying rocks by a major break in the geologic record representing all of Silurian and most of Devonian time. Late in the Devonian Period, the second sequence of Paleozoic rocks was deposited in the Black Hills region. These rocks are dominated by carbonate material, namely dolostone and limestone. Deposition of this carbonate material continued from Devonian time through the Mississippian Period.

The final group of Paleozoic rocks is another sequence of dominantly clastic silicate rocks ranging in age from Pennsylvanian to Permian. This sequence is separated from the underlying middle-Paleozoic carbonates by a minor unconformity in some areas but in other areas by no depositional break at all. Rather, the demarcation of these two sequences is based on the change from limestone to sandstone. This latest Paleozoic sequence is characterized by quite variable lithologies consisting of nearly pure quartz sandstone, calcareous sandstone, gray shale, red shale, and nearly pure limestone. The uppermost limestone in this sequence, the Minnekahta Formation, forms the outer margin of the region typically thought of as the "Black Hills." In fact, the name of the region is derived from its dark color (as viewed from a distance), the tone of the coniferous vegetation which covers the Minnekahta.

Beyond the ridge of the Minnekahta Limestone, the sedimentary sequence becomes progressively younger away from the Black Hills. With only minor exceptions, all the rocks within about a twenty-mile radius of the margin of the Hills are of the Mesozoic age and dip, at first steeply and then gently, away from the uplift.

The oldest, and innermost, of these rocks is the Spearfish Formation which consists of several hundred feet of red, silty shale with minor amounts of gypsum. The shale is highly subject to erosion and forms a broad valley typically referred to as the "Red Valley" or "Racetrack." The age of this unit is difficult to determine because it contains relatively few fossils, but it appears to be Late Permian and Early Triassic in age throughout most of its extent. It is, therefore, the earliest Mesozoic rock unit in the region.

Unconformably overlying the Spearfish Formation is a unit, composed of gypsum with minor amounts of siltstone and shale, referred to as the Gypsum Spring Formation. It, too, was deposited in a terrestrial environment. Both rock units represent deposition of sediments very near, but above, sea level. This marks the oldest major sequence of terrestrial sediments preserved in the Black Hills region.

By contrast, a sequence of marine sandstone and shale up to 700 feet thick, the Sundance Formation, overlies the Gypsum Spring Formation. Locally, the Sundance Formation contains numerous fossils of marine organisms which clearly demonstrate that the unit formed in a marine basin.

Once again, however, the Black Hills region was subjected to some uplift as shown by the succeeding units, the Unkpapa Sandstone and Morrison Formation, which are of terrestrial origin. They consist of sandstone and shale which have yielded numerous bones of dinosaurs and other terrestrial vertebrates and invertebrates. As a matter of fact, the Morrison Formation is probably the best known dinosaur-bearing rock unit in the United States.

All the rocks from the Gypsum Spring Formation through the Morrison Formation have been dated as Jurassic in age. This interval of time may be thought of as one during which the Black Hills was either just above or just below sea level and during which sediments were accumulating almost continuously. The sequence of rocks above the Morrison Formation is almost entirely marine in origin. These rocks, of Cretaceous age, constitute the youngest broadly distributed rock units in the Black Hills area.

The oldest of the Cretaceous units, the Lakota and Fall River formations, are dominantly sandstone although, locally, lig-

nitic shale (the Fuson Member) or limestone (the Minnewaste Member) may be present.

Above the Fall River Formation, most of the remainder of the Cretaceous sequence consists of gray shales with relatively minor sandstone and limestone, all of marine origin. These rock units are of considerable importance because they have a total thickness of about 4000 feet and cover much of the land surface in South Dakota west of the Missouri River as well as equally large areas in adjacent states.

The end of the Cretaceous Period was marked by deposition of the Fox Hills Sandstone and the overlying Hell Creek Formation. The Fox Hills is a sand body that represents the end of marine conditions in western South Dakota and eastern Wyoming, whereas the Hell Creek Formation represents the beginning of another cycle of deposition of nonmarine beds. Deposition of this type continued almost uninterrupted until the most recent part of the history of this area.

Cenozoic sedimentary rocks in the immediate vicinity of the Black Hills are limited in scope and difficult to identify. They consist primarily of gravel and sand deposited in valleys formed by streams issuing from the Black Hills during the Pleistocene, the youngest epoch of the Cenozoic. In areas adjacent to the Black Hills, however, these rocks become much more widespread and are readily recognizable. The Badlands region of South Dakota, for example, is formed in rocks of Cenozoic age. Similar sediments deposited in the Black Hills have been removed from most of their former extent by erosion. Other Cenozoic rocks in the area include numerous igneous intrusions, such as that exposed on Bear Butte, which are confined to a zone about 10 miles wide extending across the northern Hills. Finally, small amounts of Cenozoic volcanic rocks occur in the north-central part of the Black Hills.

STRUCTURAL FRAMEWORK

As described in the foregoing section, the bulk of the rocks in the Black Hills region are distributed in an orderly fashion; that is, they become progressively younger outward from the center of the Black Hills Uplift (fig. 2.1). This orderly distribution of rocks is a result of the major structure that forms the region. The dominant structure is a broad dome about 120 miles long in a north-south direction and 60 miles wide in an east-west direction. The dome is a result of the uplift of at least two blocks of Precambrian rock, apparently forced vertically more or less like huge pistons. The eastern block underlies nearly all of the Black Hills in South Dakota and was subject to the greatest amount of uplift (fig. 2.2). It is separated from the western block along the Wyoming-South Dakota border by a monocline. The western block moved upward less than did the eastern block, causing the Black Hills dome to be somewhat asymmetrical, that is, gently inclined on most of the western margin and steeply inclined on the east. This control of the Black Hills Uplift by vertical movement of rocks at great depth formed the major structure in the area and, indeed, has been a dominant structural element here throughout much of geologic time. Comparison of the rocks in the Black Hills region with rocks in the basins east

and west of the Uplift indicates that the Black Hills have been a positive region throughout much of post-Precambrian time. The term positive, in this context, refers to the fact that the area has been at, or above, the level of the surrounding area through most of this long time interval. Rocks of the same age in the Williston Basin to the east and northeast of the Black Hills, and in the Powder River Basin to the west of the Black Hills, tend to be much thicker and to constitute a more complete record of geologic time than do the rocks in the Hills. This thinning of sedimentary rocks over the Black Hills strongly suggests that the Precambrian rocks beneath the Hills were mobile early in the Paleozoic Era and suffered vertical movement periodically throughout most of subsequent geologic time, not just at the end of the Cretaceous Period when the final major uplift occurred.

This major uplift, referred to as the Laramide Orogeny, was the dominant mountain-building episode, not only in the Black Hills, but also in the Bighorn Mountains and the Rocky Mountains farther to the west. It was at this time that most of the uplift occurred in the region, and the Black Hills began to take on their modern form. Subsequent to that time the only structural activity of note was local doming due to intrusion of igneous bodies throughout the northern part of the Black Hills. This resulted in such prominent features as Bear Butte, the Lead Dome, Crook Mountain, and Sundance Peak.

The broad domal structure of the Black Hills can be demonstrated readily by observing the attitude of the rocks as one enters the Black Hills. On virtually any high-

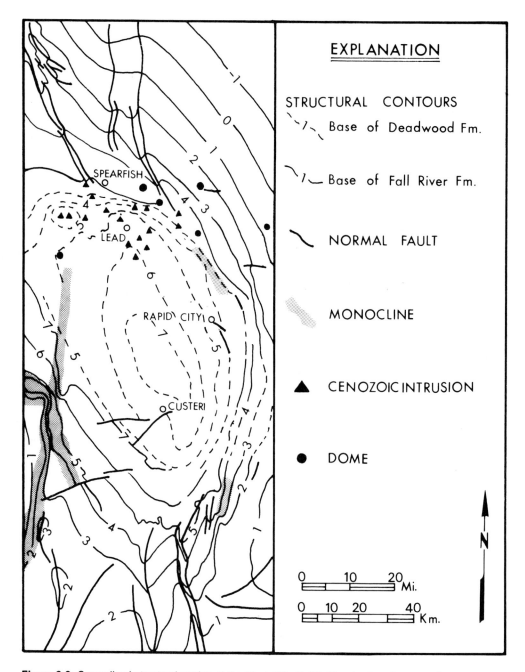

Figure 2.2. Generalized structural setting of the Black Hills Uplift. Structural contours are lines connecting all points of equal elevation on some geologic horizon, in this case either the base of the Deadwood Formation or the base of the Fall River Formation. Elevations are given in thousands of feet above mean sea level. The concentric pattern, with highest elevations near the center of the Black Hills, clearly portrays the domal structure (modified from Lisenbee 1975).

way leading into the Hills, the rocks dip or are inclined away from the Hills. By carefully observing the attitude of the rocks it is possible not only to define the major domal structure, but also to recognize the presence of more local structures which are secondary to the broad doming. Many of these minor structures are evident in the Black Hills region and appear to result either from faulting and minor folding associated with the Laramide Uplift or are structures associated with igneous activity. Where relevant, these structures will be discussed and described at specific localities covered in the guide.

GEOMORPHOLOGY

Overview

The geomorphic expression of any area is controlled by its climate, the nature of the underlying rocks, and the geologic structure of the rocks. If these three factors are well known, it is possible to "model" the history of development of an area with moderate accuracy. When this is done it can be demonstrated that the area will not take on a single type of appearance which it will maintain throughout the rest of earth history but, instead, will continually undergo changes as new landforms develop and old ones are obliterated. Therefore, an area can be thought of as being subjected to a kind of aging process involving alteration of the land surface due to weathering, erosion, and deposition. No matter how these processes act, however, they can only modify the expression of the underlying rocks and structures.

In the Black Hills, the climate is semi-arid to semi-humid. The region receives 10-20 inches of rainfall per year, slightly higher at higher elevations; and, as with most of the upper midcontinent region, most of the rainfall occurs in relatively short, often violent, bursts separated from one another by long periods of little or no rainfall. Temperature fluctuations are about as extreme as anywhere in the United States, approaching —40°F. in the winter and exceeding 100°F. in the summer. Characteristic of this type of climate are relatively sharp and angular landforms resulting from the dominance of physical weathering processes of abrasion, freezing, and thawing. These contrast with better-rounded landforms which develop in more humid regions where chemical solution of mineral grains occurs. Similarly, in arid and semi-arid regions limestone and other types of carbonate rocks tend to be generally resistant and, therefore, form ledges and cliffs. The same rocks are far more susceptible to the chemical weathering typical of humid regions where they are not generally as prominent as ridge-formers. Clearly, each rock type has its own specific set of properties relative to weathering and erosion, and these properties make rock units distinguishable from one another even though there are only subtle differences in lithology. Furthermore, the expression of any rock unit will differ depending upon its topographic position, that is, whether it is exposed on the side of a steep-walled cliff or at the surface in a broad, relatively flat area.

The major geomorphic features in the Black Hills are arranged either in a concentric pattern paralleling the outline of the Black Hills (fig. 2.3) or are radial to it; that is, they radiate out from the core of the Hills toward the margins. Because of the domal structure which dominates the region, geomorphic features associated with the rock units are typically expressed in a

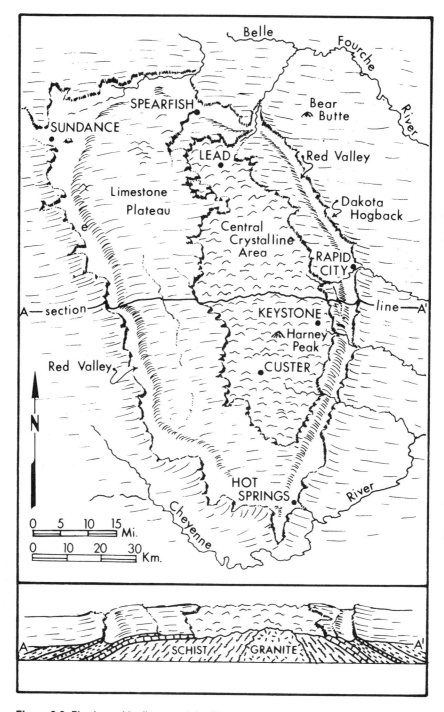

Figure 2.3. Physiographic diagram of the Black Hills showing the relationship between the broad domal structure and the erosional expression (modified from Thornbury 1954).

concentric fashion. Those geomorphic features most intimately associated with stream processes are arranged in radial fashion within the Hills and in concentric fashion beyond the margin of the Hills (fig. 2.4). Examination of these broad geomorphic features can be observed best by traveling to one of many overlooks such as Terry Peak, Harney Peak, Dinosaur Overlook, or any similar promontory where a broad area can be viewed.

Precambrian Core

The rocks in the core of the Black Hills, the Central Crystalline area (fig. 2.3), can be subdivided generally into two broad areas, the northern and peripheral region composed of metamorphic rocks and a south-central region of granitic rocks. The metamorphic and the granitic regions are uniformly resistant to weathering which results in development of a generally dendritic (branching) drainage pattern except where the streams are locally controlled by jointing, minor fracturing, or foliation in the rock. Foliations locally produce steep-walled, smooth, often shiny surfaces which impart a "grain" in stream patterns. Occasionally quartzite beds, which are more resistant to erosion than the surrounding phyllite and schist, are also responsible for defining stream orientation. Stream patterns produced by the foliation and quartzite units are, however, relatively minor and of small scale. In the granitic terrain the major features producing grain are joints (fractures) in the granitic rock. Where they occur, such as in The Needles area, long linear topographic features result from weathering and erosion along the joint planes.

Some of these features are truly spectacular illustrations of the relationship between weathering and jointing. At the interface between the granitic and the metamorphic terrains in the southern Black Hills, granitic pegmatites commonly intrude the metamorphic host rock. Where intrusion occurs the pegmatites tend to weather more slowly than the surrounding metasedimentary rocks and, therefore, stand out in relief.

Paleozoic Terrain

A much more obvious "grain" is developed in the Paleozoic terrain than in the core of the Black Hills. This is particularly clear when viewed from a promontory within the Black Hills, such as Terry Peak, but can be observed also by traveling on almost any of the highways into the Black Hills. When one does so, it is apparent that the Deadwood, Pahasapa, Minnelusa, and Minnekahta formations tend to produce prominent ridges or escarpments which are separated from adjacent ridges by intervening less resistant rock. The Pahasapa escarpment is the most outstanding of these. The doming of the central Black Hills is asymmetrical, that is, it is generally somewhat steeper on the east flank than on the west. This asymmetry has resulted in greater dissection or stripping away of the Paleozoic cover from the east margin than from the west. Therefore, the western half of the Black Hills is covered on the upland surface by the Pahasapa Formation which produces a relatively broad gently-sloping surface called the Limestone Plateau (fig. 2.3). This plateau is cut by a few streams such as Spearfish Creek. On the eastern flank, on the other hand, the core is ex-

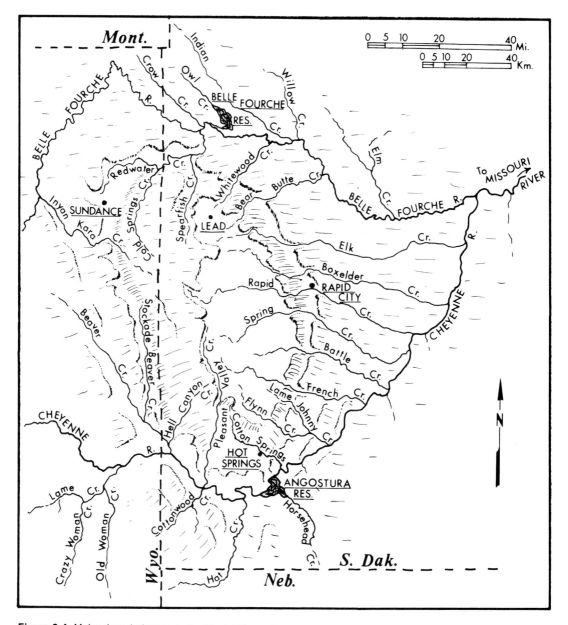

Figure 2.4. Major river drainages in the Black Hills region.

posed and one must proceed toward the margin of the Black Hills before Paleozoic rocks are encountered. In these areas, the resistant rock units stand as very long linear ridges which result from the dip of the beds. For example, the beds in the southernmost part of the Black Hills dip very gently to the south; the development of escarpments in this area is much reduced compared to that on the north and northeastern flanks of the Hills where the dips are somewhat steeper (20-30°) and the escarpments are much more prominent.

Another geomorphic feature of interest in the Paleozoic terrain is the development of caves in the Pahasapa Formation. Cavern development is relatively common in this rock unit and, in the same unit, one can observe other features associated with solution of limestone. Disappearing streams and springs are relatively common. When the rock is examined closely, small cavities, called vugs, can be observed lined with perfectly formed calcite crystals resulting from solution of the limestone and re-precipitation of the calcium carbonate in voids within the rock. The formation of caves in this rock is simply a larger-scale solution phenomenon related to the development of these vugs.

Because the Paleozoic rocks form steep escarpments in many parts of the Black Hills and because they exhibit jointing and some faulting associated with the Laramide Uplift, they are subject to landsliding. On almost any of the highways leading into the Black Hills the effects of the landsliding can be observed as slopes stripped of vegetation, piles of rock debris at the base of the slopes, and freshly repaired sections of road.

The Racetrack

Surrounding the Paleozoic terrain of the Black Hills is a broad, generally flat area underlain by red shale of the Spearfish Formation. This feature, the Red Valley or Racetrack (fig. 2.3), is extremely obvious around most of the margin of the Black Hills—so much so that it was recognized and named by the Indians long before the area was scrutinized by geologists. The Red Valley area takes on its appearance because the Spearfish Formation is relatively weak and is readily eroded and because it is located stratigraphically between resistant rock units, the Minnekahta Formation below and the Sundance, Lakota, and Fall River formations above. These units form escarpments separated from one another by the Racetrack. An additional point of interest is that relatively few streams parallel the Racetrack. Rather, the major drainages flow directly across the valley and through the escarpments on the outer margin of the hills (fig. 2.4). One might expect that such a valley would contain streams paralleling the area of outcrop of the Spearfish Formation, but this is not the case. This observation suggests that the streams radiating from the center of the Black Hills and ultimately flowing into the two major concentric drainages, the Belle Fourche River and the Cheyenne River, were formed originally long before the Racetrack developed. The streams flowed radially from the Black Hills Uplift out onto the Great Plains and cut down through the Mesozoic and, finally, the Paleozoic rocks. By the time the Racetrack began to form as a distinct geomorphic feature, the streams had already formed well-defined valleys cutting directly across the flanking structures of the Black Hills. That general drainage pattern continues to the present time.

Dakota Hogback

In most areas surrounding the outer margin of the Racetrack, prominent escarpments of sandstone rim the Black Hills. These escarpments are composed of and held up by the Sundance, Lakota, and Fall River formations. The latter two form an escarpment which is typically referred to as the Dakota Hogback (fig. 2.3), a name taken from correlative rocks in other areas of the upper midcontinent. This escarpment typically is thought of as the outer boundary of the Black Hills, and it generally marks the outer limit where beds can be observed readily dipping away from the Black Hills Uplift. Of particular interest is the fact that this Dakota Hogback forms one of the recharge areas for groundwater. Water derived from percolation of surface water into the Lakota and related rock units flows through these rocks and, by drilling, provides a groundwater resource of considerable importance over much of the area of South Dakota, North Dakota, and part of Wyoming. Beyond the Dakota Hogback the terrain is dominated by the Late Cretaceous dark shales. Here the topography is generally subdued and any major relief is related primarily to dissection along the Cheyenne and Belle Fourche rivers and their tributaries. The Cretaceous shales tend to weather into gently rounded surfaces which support only meager vegetation.

Igneous Intrusions

The final geomorphic features of note are the peaks and domes underlain by igneous intrusions in the northern Black Hills.

These features occur in a belt extending from Bear Butte westward across the Hills. They can be recognized by the fact that, in almost all cases, they appear to have a topographic expression quite different from that of the surrounding sedimentary terrain. Where exposed, the cores of these intrusions are highly resistant to erosion and they stand high as at Bear Butte and Sundance Mountain. Where not exposed by erosion they have, nevertheless, domed up the rocks around them so that small concentric bands of escarpments can be observed as at Elkhorn Peak, west of Whitewood, South Dakota.

GEOLOGIC HISTORY

General

Rocks exposed in the area of the Black Hills Uplift record events in earth history extending from the Recent back at least 2.5 billion years. Therefore, the record encompasses over half of all of earth history. In fact, very few areas around the world give evidence of geologic history older than about 2.5 billion years. As one might expect, the geologic record in the Black Hills generally becomes more complete and more detailed as one progresses up the geologic column toward progressively more recent intervals of time. This is because younger rocks are generally better exposed, areally more extensive, and tend to have undergone less alteration by geologic processes.

Precambrian Geology

The oldest rocks exposed in the Black Hills area are those which formed during the Precambrian interval of geologic time. A map of the region (fig. 2.5) shows that less than 10% of these rocks are of igneous

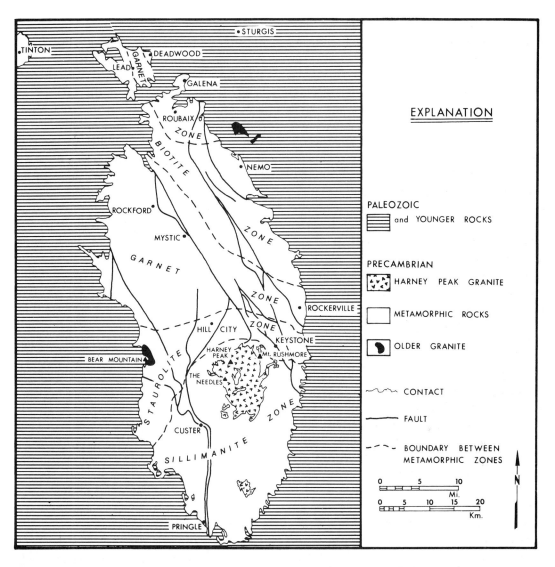

Figure 2.5. Major Precambrian geologic features in the Black Hills (modified after Redden 1975).

origin, having formed by the intrusion and crystallization of magma, or molten rock material, at great depth. The majority of Precambrian rocks in the Black Hills are of metamorphic origin; they owe their present mineral composition, textures, and structures to recrystallization resulting from application of heat and pressure at considerable depth after they were originally formed as other kinds of rocks. Although the gross aspects of the Precambrian geology in the Black Hills are now understood, many of the details have yet to be unraveled. Those who have studied these rocks suggest that they have been deformed at least three times and have been subjected to two intervals of metamorphism as well as to two episodes of intrusive igneous activity (Redden and Norton 1975).

The oldest of the Precambrian igneous rocks are small bodies of granite which are exposed at Bear Mountain and in an area approximately four miles north of Nemo (fig. 2.5). At both localities the rocks are composed of the following minerals: quartz, plagioclase feldspar, microcline feldspar, and micas (biotite and/or muscovite), all commonly identifiable without a hand lens. Locally well-developed gneissosity, or metamorphic layering, and zones of crushing testify to the antiquity of these granites and indicate that they have been deformed since they were originally intruded. Age determinations, based on the natural radioactivity of minerals in the granites, indicate their formation about 2.5 billion years ago (Ratté and Zartman 1970; Zartman and Stern 1967).

The more common Precambrian igneous rock in the region is known as the Harney Peak Granite, found exclusively in the south-central part of the Black Hills (fig. 2.5). This rock is well exposed at the Mount Rushmore Memorial, Harney Peak, Sylvan Lake, and The Needles localities. Although shown as a single large area on the map, the Harney Peak Granite occurs as numerous small bodies scattered through the southern Black Hills. Many of these bodies were intruded as sills—tabular sheets whose boundaries are parallel to layering in the adjacent non-igneous rock (fig. 2.6). Other sheetlike bodies, referred to as dikes, cut across the layering of the adjacent rocks. Approximately 80% of the area shown as Harney Peak Granite in figure 2.5 actually consists of granitic rock. Particularly along the periphery of the area shown, the rock is a mixture of granitic dikes and sills with non-igneous rock (fig. 2.7).

The granitic rock is light gray to pink and is composed largely of the minerals albite (a plagioclase feldspar), quartz, microcline, muscovite, and tourmaline. Grain size varies widely, crystals ranging from fractions of an inch to several feet in maximum dimension. Although much of the granitic rock consists of grains 0.5 to 1 inch in diameter, locally microcline crystals, up to 1 foot in diameter, occur scattered through the finer-grained matrix. Much of the rock mapped as Harney Peak Granite is actually granitic pegmatite, a rock containing the same minerals as granite but very coarse-grained. Many of the granitic pegmatites are of commercial value and are discussed in a later section. Commonly the Harney Peak Granite has a layering consisting of alternating, vaguely defined very coarse-grained plagioclase-microcline-quartz zones and finer-grained plagioclase-quartz zones. The granite is typically transected by several sets of joints, or parallel fractures, which define tabular slabs as at Sylvan Lake (fig. 2.8) and distinctive pinnacles of granite such as those

Figure 2.6. Thin granitic pegmatite sills which have intruded schist southeast of the Norbeck Memorial on Route 16A. Note parallelism of the sill contacts with the steeply inclined platy structure in the schist.

Figure 2.7. Large irregular body and smaller bodies of Harney Peak Granite which have intruded schist near approach to Mount Rushmore on Route 244.

Figure 2.8. Well-developed jointing (parallel fractures) in the Harney Peak Granite at Sylvan Lake.

well displayed at The Needles. Locally, the granite contains inclusions of non-igneous rock engulfed in the magma during its intrusion.

Based on its radioactivity, the Harney Peak Granite formed 1.74 billion years ago (Riley 1970) and is thus distinctly younger than the granites referred to earlier. Redden and Norton (1975) believe that the granite formed at considerable depth in the earth's crust, approximately ten miles below the surface. Subsequent erosion during Precambrian time exposed it to the surface prior to its burial by Cambrian sediments.

The rocks into which the Harney Peak and older granites were intruded consist of a diverse variety of metamorphic rock types. Based primarily on a comparison of their chemical composition with that of various kinds of non-metamorphic (sedimentary and igneous) rocks, we can infer the nature of the rocks which, prior to their metamorphism, were deposited in this region. Table 2.1 lists the main kinds of Precambrian metamorphic rocks in the Black Hills as well as the probable rock type from which each was derived. In addition to gross chemical comparisons, clues as to the nature of the pre-metamorphic rocks are provided by the presence of original structures and textures which survived the metamorphism. The major non-igneous rocks originally deposited in this region, as Table 2.1 shows, were largely sedimentary rocks including shale (composed of very fine-grained clay minerals) and a wide variety of sandstones and con-

TABLE 2.1. Common Precambrian metamorphic rocks and the rocks from which they were derived.

Metamorphic Rock Type	Probable Original Rock Type
Quartzite	Quartz Sandstone
Metaconglomerate	Quartz Conglomerate
Slate	Shale
Phyllite	Shale
Mica ⎫ Garnet ⎬ Schist Staurolite ⎪ Sillimanite ⎭	Shale
Graphitic Schist	Carbonaceous Shale
Quartz-Mica Schist	Argillaceous Sandstone
Metagraywacke	Graywacke Sandstone
Meta-arkose	Arkosic Sandstone
Dolomitic Marble	Dolostone
Amphibolite	Dolerite/Gabbro Intrusion
Amphibolite	Basalt Lava Flow
Hornblende Schist	Basalt Lava Flow
Metabasalt	Basalt Lava Flow
Meta-Iron Formation	Iron Formation

glomerates (composed of larger grains of quartz and feldspars in various ratios). In addition, some of the original rocks consisted of basaltic lava flows interbedded with the sedimentary rocks, as well as intrusions of dolerite or gabbro (rocks coarser grained than basalt but also composed largely of the minerals plagioclase and pyroxene).

These rocks, which are thought to have a total thickness of 60,000 feet, were deformed and metamorphosed largely prior to intrusion of the Harney Peak Granite (Redden and Norton 1975). Variation in the intensity of metamorphism is reflected by the distribution of certain key minerals in the recrystallized rocks. The progressive appearance of the minerals biotite, garnet, staurolite, and sillimanite reflect conditions of increasing metamorphic intensity in that order. Zones in which these minerals occur are shown in figure 2.5. Of interest is the fact that at least some of the zone boundaries bend around the main mass of Harney Peak Granite, suggesting that intrusion of the granite was, in part, a cause of the metamorphism of these rocks (Redden 1968; Ratté and Wayland 1969). During metamorphism of the deeply buried shales, sandstones, and associated igneous rocks, some minerals were destroyed and a series of new minerals formed in their place, accompanied by recrystallization and a general increase in grain size. The widespread shales were converted to schists possessing a distinctive planar structure referred to as schistosity. This feature allows the rock to be split relatively easily into thin slices because the grains of mica minerals, which are prevalent in schists, are oriented parallel to one another.

The schistosity in these rocks is inclined steeply, vertically in some cases, throughout much of the Precambrian area. As granitic bodies are approached, rock units as well as the inclination of the schistosity locally bends around each mass of granite such that individual dome-like features are defined by this structure of the metamorphic rocks. The largest dome is that associated with the Harney Peak Granite (Redden and Norton 1975). Other relatively large domes are associated with older granites such as those near Bear Mountain.

In areas where the intensity of metamorphism was relatively low, the same shales were converted to slate or phyllite rather than schists. Thus the resulting variety of metamorphic rocks results from two conditions: (1) the intensity of metamorphic recrystallization, and (2) the nature of the original rock. As Table 2.1 shows, quartz sandstone became quartzite, dolostone became dolomitic marble, and basalt lava flows were converted to amphibolite or hornblende schist. At least some of the original rocks may have consisted of iron formation composed of intimately-interbedded iron-rich layers and silica-rich layers. The metamorphic equivalent of iron formation occurs within the Black Hills along with several other rock types listed in Table 2.1.

As figure 2.5 shows, the Precambrian metamorphic rocks have been transected by a series of major faults which trend northerly or northwesterly. These are zones along which the rocks on one side have shifted past the rocks on the other side.

Phanerozoic Geology

All of geologic time, from the end of the Precambrian to the Recent, is referred to as the Phanerozoic interval. The events of this time interval of about 600 million years are recorded primarily in sedimentary rocks in contrast to the dominantly metamorphic and igneous rocks of the Precambrian interval.

The oldest Phanerozoic rocks deposited in the Black Hills region are embraced in the Deadwood Formation of Late Cambrian age (fig. 2.9). This rock unit, deposited starting about 500 million years ago, helps bracket the erosional interval that affected the region. It is worth noting, for example, that the gap in the Black Hills rock record exceeds one billion years which is twice the amount of time encompassed in all of earth history from the beginning of deposition of the Deadwood Formation to the present time. As a matter of fact, it is amazing to examine the contact between the Precambrian rocks and the Deadwood Formation in the Black Hills and to consider that the interval of time represented by that erosional interface embraces 25% of all of the history of the earth. It is now represented only by a relatively low-relief surface on Precambrian rocks which apparently developed in an area of humid climate where chemical weathering was rather complete and where ancient mountains of Precambrian age were reduced to gentle hills.

In the Cambrian Period the area of the Black Hills, as with much of the midcontinent, subsided below sea level, and marine sediments were deposited across the area (fig. 2.10A). Initially, this sequence of rocks was composed of coarse,

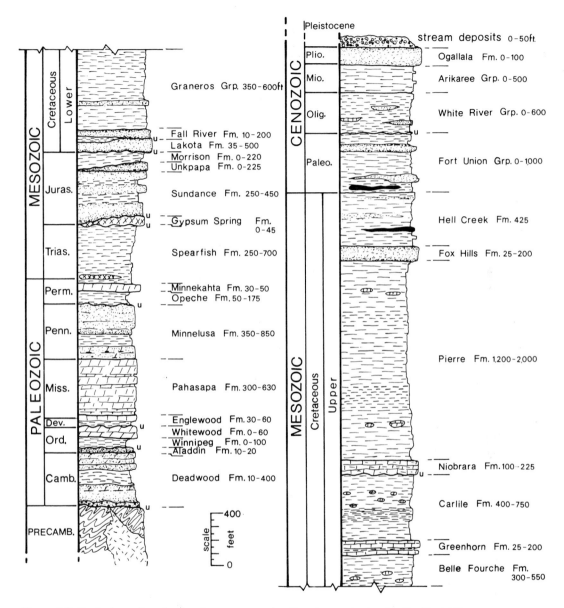

Figure 2.9. Geologic section showing the sequence of sedimentary rocks forming the Black Hills.

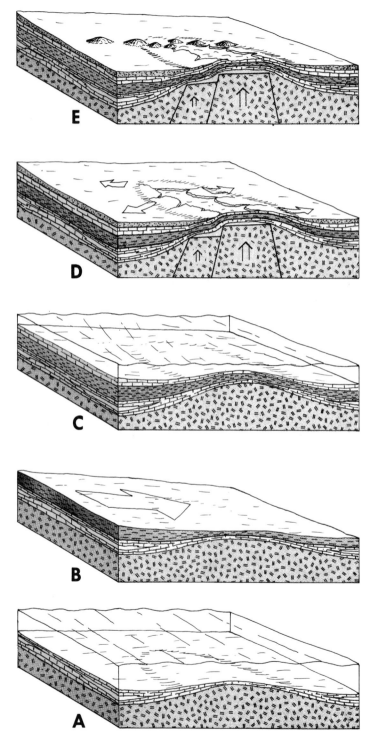

Figure 2.10. Summary of structural and sedimentational history of the Black Hills. A, deposition of marine sediments throughout much of the Paleozoic. B, deposition of terrestrial sediments during early Mesozoic time. C, marine deposition during late Mesozoic time. D, onset of major Black Hills uplift near the end of the Mesozoic. E, intrusion of igneous rocks in the northern Black Hills in early Cenozoic time. Continued erosion has resulted in formation of the present landscape depicted in figure 2.3.

clastic material, conglomerate in some places and sandstone elsewhere, which graded upward into finer Ordovician clastic material and finally into carbonate rocks. Determination of the age of these rocks is greatly enhanced by the fact that they possess a good, though not abundant, fossil record. Organisms in the Deadwood Formation are restricted almost entirely to fragments of trilobites and to tracks, trails, and burrows of organisms, perhaps soft-bodied, which have left no other physical evidence of their existence. Recently fish remains have been found in the Deadwood Formation near Sundance, Wyoming. These represent the oldest remains of vertebrate fossils known anywhere in the world. In the Ordovician sequence, notably in the Whitewood Formation, the fossil record is somewhat more diverse and certainly more abundant. Snails, cephalopods, and corals can be found in almost any exposure of rock. These organisms, typical of Ordovician rocks throughout the area, are a clear indication of a marine habitat insofar as their closest relatives living today occupy only a marine environment.

Following deposition of this early Paleozoic sequence, the Black Hills region was apparently elevated somewhat above sea level because no rocks of Silurian or Early Devonian age are known. Marine rocks of the Silurian and Early Devonian age are found in the Powder River Basin and the Williston Basin flanking the Black Hills, but none are known over the area of the Black Hills Uplift. This suggests that either marine rocks were deposited over the Black Hills during this interval of time and were eroded prior to deposition of the Late Devonian-Mississippian rocks, or that the area was positive throughout this interval of time and no rocks of these ages were

deposited. In any event, the next sequence of sedimentary rocks observed above the Whitewood Formation are of latest Devonian and Mississippian age and are embraced in the Englewood and Pahasapa Formations. Although a relatively thin sequence of calcareous shale marks the lower part of the Englewood Formation, the two units are dominantly carbonate rock units. Examination of these rocks in detail indicates that they were deposited as fragments of carbonate material ranging in grain size from silt to coarse sand and pebbles. The finer material, deposited as a carbonate mud, produces a dense, generally structureless rock whereas the rocks formed of larger grains of carbonate material are typically "hashes" of broken fossils (remains of animals living in the sea in which the rock units were deposited). Whole or nearly whole specimens of brachiopods and corals can be found in these rocks, and they enable one to date the units and demonstrate the gap in the geologic record which exists between the Whitewood and the Englewood Formations.

Near the top of the Pahasapa Formation, fossils become somewhat more abundant. Notable among them are large, dome-shaped masses of colonial coral which appear to be small patch reefs surrounded by fragments of organic material deposited in a wave-agitated, high-energy environment. This suggests that at the close of Pahasapa time the area was at, or very near, sea level.

The overlying rocks of the Minnelusa Formation bear an interesting relationship to the Pahasapa Formation. Some areas of the Black Hills show evidence of the upper Pahasapa Formation having been subjected to erosion and partial solution. In other areas, however, the rocks of the Minnelusa and Pahasapa formations appear to intertongue; that is, the lithology of the Pahasapa Formation is abruptly terminated by deposition of Minnelusa-type sediments which are in turn abruptly followed by more Pahasapa sediment and so forth. This intertonguing relationship suggests that the carbonate bank on which the Pahasapa Formation was deposited periodically underwent flooding by clastic silicate sediments, which were probably transported into the area by streams draining an adjacent silicate terrain. The area was finally flooded by this silicate material, and sediments of the Minnelusa Formation accumulated.

Deposition of marine sedimentary rocks was more or less continuous until near the close of the Paleozoic Era. Minnelusa, Opeche, and Minnekahta sediments were deposited, one above the other, with little apparent break in the sequence. Whereas the Minnelusa Formation is comprised generally of sand-sized sedimentary rock, the Opeche is characterized by finer-grained sediments (silt and clay) and the Minnekahta by carbonate material. Fossils in these three units are extremely rare although occasional invertebrates and a few vertebrate remains (e.g. shark teeth) have been found in the Minnelusa Formation.

Throughout the Paleozoic Era, the area of the Black Hills was either slightly below sea level and receiving a mantle of shallow-water marine sediments, or was elevated slightly above sea level and was being subjected to minor amounts of erosion. This picture changes markedly in the early Mesozoic Era when sediments of the Spearfish Formation were deposited. There is no clear indication that Spearfish sediments

are of marine origin. Rather, they consist of a sequence of fine-grained red sand, silt and clay, with some associated gypsum, that is suggestive of a low-relief, perhaps arid, terrestrial environment (fig. 2.10B). Similarly, the overlying Gypsum Spring Formation is suggestive of a terrestrial, perhaps marginal marine, environment. The Gypsum Spring Formation is Middle Jurassic in age and is separated from the Spearfish Formation by an unconformity, or gap in the record.

Above the Gypsum Spring Formation, the rock record becomes progressively more marine, that is, most of the rock units through Late Jurassic and Cretaceous time are of marine origin. The Sundance Formation, with a maximum thickness of over 350 feet, consists almost exclusively of sandstone and shale deposited in a marine environment (fig. 2.10C). Fossils are not common throughout most of the Sundance Formation, but in the upper part large accumulations of belemnites (cigar-shaped calcareous rods) do occur (fig. 2.11). These are the remains of squid-like cephalopods which apparently lived, died, and were preserved in the Sundance sea in enormous numbers. Locally, they accumulated in such vast numbers that exposures are nearly covered with their remains. In association with these belemnites are oyster-like bivalves and other shallow-water marine invertebrates. Above the Sundance Formation and representative of latest Jurassic time, the Morrison Formation was deposited in a terrestrial environment. Dinosaur remains have been found throughout the Morrison Formation.

From the close of the Jurassic Period to near the end of the Cretaceous Period, a sequence of shale, sandstone, and limestone

Figure 2.11. Belemnites preserved in a thin zone (just below the pencil) in the Sundance Formation.

accumulated to a thickness of nearly one mile. These rocks, almost exclusively of marine origin, mark the last major time interval during which marine sediment accumulated in the Black Hills region. These rocks locally are quite fossiliferous and contain representatives of bivalves, gastropods, cephalopods, and minor numbers of almost all other phyla of marine organisms. The seaway in which these rocks were deposited extended all the way along the front of the Rocky Mountain region. The rocks are the youngest that were deformed in the Laramide Orogeny, the mountain-building episode during which the Rocky Mountains and the Black Hills were finally uplifted.

This orogeny began in Late Cretaceous and continued into Early Cenozoic time (fig. 2.10D). As it occurred the Black Hills were uplifted, perhaps as much as two miles above sea level, and then began to erode. Sediments from the erosion of the Black Hills (and the Rocky Mountains for that matter) were deposited in the basins surrounding the uplifts. The oldest of these rocks, the Hell Creek Formation, is the last major dinosaur-bearing rock unit in North America. Early Cenozoic rocks of the overlying Fort Union Group consist of rocks quite similar to the Hell Creek rocks but contain no dinosaur remains. Deposition of non-marine sedimentary rocks derived from the erosion of these uplifts took place almost continuously from the end of Cretaceous through Pliocene time, although erosional breaks, and therefore removal of some of the material, separate many of the rock units.

Unraveling the Cenozoic geologic history of the Black Hills region is difficult. Over much of the immediate area of the Hills, erosion was clearly the dominant process; however, numerous occurrences of sand and gravel throughout the area indicate deposition of sediments as well. Most do not contain fossil material suitable for determining the age of the gravels. It is possible that some of the gravels are Oligocene in age and some are probably Pleistocene, but a great deal more work will be necessary before the sedimentary history of this interval of time can be detailed.

The northernmost portion of the Black Hills is characterized by a distinctive group of igneous bodies which were intruded during the Cenozoic interval of geologic time. These rocks are confined to a generally east-west belt approximately ten miles wide (fig. 2.12). According to Redden (1975), there are four major intrusive centers within this belt. Within the area shown in figure 2.12, one such center is at Tinton, a second is the Lead-Deadwood area, and a third is the area west and east of Galena. Individual bodies of these rocks at Terry Peak, west of Lead on Mount Theodore Roosevelt, and at Bear Butte are described later. Similar bodies are well-exposed in the Homestake Mine open cut in Lead.

The Cenozoic igneous bodies consist of both equigranular and porphyritic varieties of the following rock types: quartz monzonite, monzonite, latite, rhyolite, and phonolite. Several other rock types account for a small percentage of these bodies as well. The rocks with porphyritic texture represent typically two stages of cooling. As local pockets or reservoirs of magma cooled slowly at depth, the relatively large crystals (called phenocrysts) formed under quiescent conditions. At a slightly later time these magmas, containing previously-crystallized phenocrysts, were suddenly intruded to shallow levels in the crust. This facilitated rapid cooling of the liquid magma, within which the phenocrysts were suspended, causing this material to crystallize as a relatively fine-grained matrix around the larger phenocrysts. Depending upon the particular rock type of which a given intrusion is composed, the phenocrysts may include one or more of the following minerals: quartz, orthoclase feldspar, plagioclase feldspar, and aegirine (a variety of pyroxene).

Because these rocks are geologically young, they have intruded most of the Paleozoic and Mesozoic sedimentary rocks as well as the Precambrian rocks in the Black Hills. They occur typically as dikes, sills, cylindrical plugs, stocks, and laccoliths (fig. 2.13). They range in size from dikes and sills several feet thick to laccoliths several hundred feet thick and several miles in diameter. With the exception of the dikes, these Cenozoic bodies have planar contacts which are parallel to the orientation of schistosity in the enclosing Precambrian rocks or parallel to the orientation of bedding in the adjacent sedimentary rocks (fig. 2.14). Intrusion of the larger bodies has caused the formation of domes, particularly in the Paleozoic and Mesozoic sedimentary rocks. In several areas the igneous rocks are not exposed at the surface, but the sedimentary rocks are domed. This indicates that Cenozoic intrusions are concealed at some depth (Redden 1975). Noble (1952) presented convincing evidence that the entire Lead-Deadwood dome resulted from intrusion of Cenozoic rocks below.

Noble (1952) estimated that the depth of final intrusion of these bodies was probably less than 8,000 feet, and thus was relatively shallow when compared with the depth of intrusion of the Harney Peak Granite. He noted further that, because contact metamorphism is extremely limited in all kinds of rocks adjacent to the intrusions, most of the intruding magmas were relatively low in volatile content, were probably partly crystallized already and lacked excess heat. Although there is some textural variation among these rocks, many are closely related chemically as well as geographically, which suggests that they

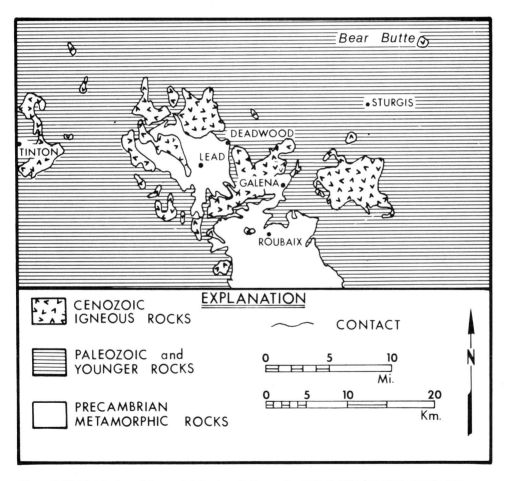

Figure 2.12. Distribution of Cenozoic intrusions in the northern Black Hills (modified after Reddin 1975).

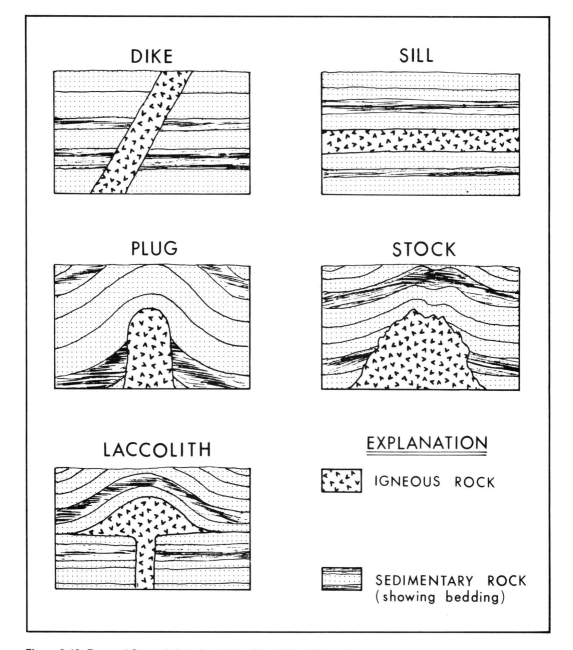

Figure 2.13. Forms of Cenozoic intrusions in the Black Hills. All sketches are cross-sections; not to scale.

were intruded at about the same time geologically and most were probably formed from the same general magma type (Darton and Paige 1925). Age determinations (summarized in Shapiro 1971) based on radioactivity of minerals in six different Cenozoic rock types from this region, confirm that these bodies were intruded within the same geologic time interval. The ages fall between 38.8 and 60.5 million years, a relatively narrow range of geologic time; this coincides with the interval referred to as the Eocene Epoch.

Although the best known and most widespread of the Cenozoic-age igneous rocks in the Black Hills are intrusive rocks, as discussed above, Darton (1912) described a small area of volcanic, or extrusive, igneous rocks of this general age. These rocks occur approximately seven miles south of Deadwood and west of Roubaix (fig. 2.12) on the property of the Tomahawk Lake Country Club. Described later in the guidebook, the rocks here consist of a variety of fragmental types as well as obsidian or volcanic glass. Recent age determinations, based on their radioactivity (Kirchner 1977), suggest that they formed approximately 10.5 million years ago during the Miocene Epoch. If correctly interpreted, these important rocks greatly expand the time range during which Cenozoic igneous activity occurred in this region and indicate that it extended to a much younger interval of time than previously thought.

Throughout much of Cenozoic time the Black Hills region was elevated well above sea level and was undergoing active ero-

Figure 2.14. Cenozoic sill-like intrusion which is slightly discordant in relation to the steeply inclined platy structure of the enclosing schist.

sion resulting in sediments being deposited around the margins of the uplift and extending out into central South Dakota and east-central Wyoming. In this same interval of time, the major present-day Black Hills physiographic features were probably defined. Pleistocene geologic history, recorded by deposits of glacial origin in many of the northern midcontinent area, is recorded only by stream deposits in the Black Hills. There is no evidence that the Black Hills were ever glaciated and no glacial deposits are known from the region. Instead, the area was probably subjected to somewhat higher rainfall during this period and was probably rather extensively eroded. Continued erosion has resulted in the final sculpting of the landscape that we see today (fig. 2.3).

When the present-day Black Hills are observed, therefore, we see an area of high relief and quite varied landforms that are the product of sediment deposition, igneous intrusion, mountain-building, and erosion that have been active in the area for 2.5 billion years. Evidence for the history of the area is recorded in the rocks, and this history can be worked out in some detail by application of relatively simple geologic concepts. Understanding this history not only helps us to better understand the development of the earth on which we live but also gives a much greater appreciation for the topographic features which are typically viewed only as objects of great beauty and recreational interest. Examination of the rocks at many of the localities described in subsequent sections of this guide will focus in more detail on aspects of the interpretation of earth history.

ECONOMIC GEOLOGY

Granitic Pegmatites

The Black Hills is a world-famous locality for a remarkable group of rocks referred to as granitic pegmatite. The rock has the mineral composition of typical granite, being dominated by quartz and feldspars. However, it is much more coarsely grained than granite. Extreme examples include crystals with an exposed length of 47 feet obtained from the Etta Mine in the early days of mining (Norton et al. 1964). Typical pegmatites contain an abundance of crystals several inches or several feet across (fig. 2.15). Because of the unusually large size of individual minerals in pegmatites and because of the abundance of commercially valuable minerals in some of them, pegmatites may be of considerable economic value; and those in the Black Hills have been mined since 1879. Although numerous pegmatites occur in the Tinton area at the northwest end of the Black Hills, the greatest concentration of them is in the southern Black Hills. Here some 20,000 pegmatites occur within an area of approximately 300 square miles surrounding the Harney Peak Granite (Norton 1975), as shown in figure 2.16. Isograms, or contours which depict the number of pegmatites per square mile, show that the concentration of these bodies increases generally toward the Harney Peak Granite. However, the greatest concentration occurs southwest of the granite due south of Custer.

Both zoned and unzoned granitic pegmatites occur in the Black Hills. However, only the zoned type are of commercial value. Some 200 such pegmatites occur in

Figure 2.15. Typical thin pegmatitic sill in road cut along U.S. Highway 16A southeast of the Norbeck Memorial. Note oriented tourmaline crystals, several inches long, near right contact across from pencil.

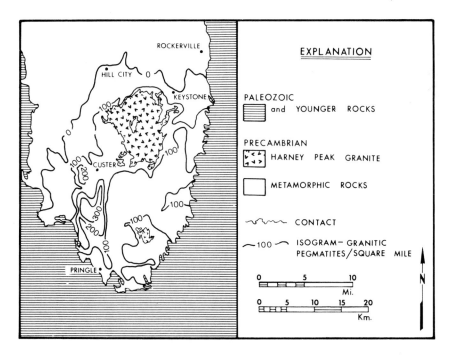

Figure 2.16. Distribution of granitic pegmatites and their relationship to the Harney Peak Granite, southern Black Hills (after Norton 1975).

the region. The commercial value of the zoned pegmatites lies in the fact that a given mineral is concentrated in a particular zone, and thus only this zone of the pegmatite must be mined to produce an economically valuable product. Moreover, the zoned pegmatites are typically coarser grained than others, and thus they are readily mined by hand picking.

In general, this kind of pegmatite possesses three types of internal structural units: zones, fracture fillings, and replacement bodies (fig. 2.17). In the Black Hills pegmatites, zones are by far more common than the other two kinds of internal units. The zones occur essentially as concentric shells which reflect the overall shape of a given pegmatite. Their thickness is largely

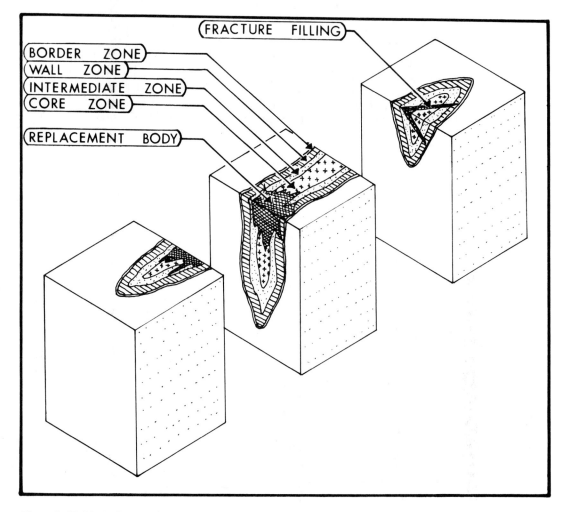

Figure 2.17. Block diagram showing types of internal structural units common to zoned granitic pegmatites (after Cameron et al. 1949).

a function of the overall size of the body. Typically, the outermost or border zone is a fraction of an inch to several inches thick. The wall zone is commonly a foot to ten feet in thickness, whereas the intermediate zone or series of zones ranges from one foot to 30 or more feet in thickness. Finally, the core zone may be as much as 40 or more feet in thickness. Typically, the zones are characterized by specific suites of minerals arranged in a definite order. For the Black Hills pegmatites, the general sequence of these suites is given in Table 2.2.

Many of the zoned pegmatites occur as lenticular dikes or sills (fig. 2.15) and as irregular stocks. They are typically lenticular in map and cross-section views, with a rounded top, kneel-like bottom (fig. 2.17), and undulations along their contact with the enclosing rock which is commonly schist (Norton 1975).

The first pegmatite mine to be developed in the Black Hills was the Crown Mine opened in 1879 near Custer. This pegmatite

TABLE 2.2. Sequence of mineral assemblages common in zoned granitic pegmatites of the Black Hills (after Redden 1963)

Assemblage	Essential minerals composing more than 5 percent of each zone	Zone			
		Border	Wall	Intermediate	Core
1	Muscovite, quartz, plagioclase	X	X	X	—
2	Quartz, plagioclase	X	X	X	X
3	Perthite, quartz muscovite, plagioclase	X	X	X	X
4	Perthite, quartz	—	—	X	X
5	Perthite, quartz plagioclase (usually cleavelandite), amblygonite, spodumene	—	—	X	X
6	Plagioclase (cleavelandite), quartz	—	—	X	X
7	Plagioclase, quartz, spodumene	—	—	X	X
8	Quartz, spodumene	—	—	X	X
9	Plagioclase, quartz, lepidolite	—	—	—	X
10	Quartz, microcline	—	—	X	X
11	Microcline, plagioclase, lithia-mica	—	—	—	X
12	Quartz	—	—	—	X

produced sheet mica which dominated the economically valuable materials extracted in the early days. Sheet mica mining was given a boost during 1942-1945 and 1952-1962 because of the demand for mica for military purposes, largely for electronic equipment. In 1898 extraction of lithium minerals was begun at the Etta Mine near Keystone. In fact, Black Hills pegmatites constituted the dominant world-wide source of lithium from 1898 until 1952. Lithium production was at a peak during World Wars I and II and during the Korean War when several lithium minerals were mined from different pegmatites. Since the Etta Mine closed in 1960, lithium mining has been negligible in the Black Hills. The Hugo Mine (fig. 2.18) supplied

the first commercial microcline feldspar in 1923 and this mineral is still mined in the Black Hills today. Although it was mined sporadically as a by-product in early years (Norton 1975), beryl became an important mineral product from these pegmatites during World War II. The Black Hills pegmatites have also produced tantalum and niobium as well as tin, white quartz, and rose quartz. A very large number of mineral specimens from these pegmatites have been sold over the years as well.

Among all of the pegmatite minerals mined in the Black Hills, potassium feldspar is the most important. The particular potassium feldspar mineral mined is referred to as microcline perthite which consists largely of microcline and very thin

Figure 2.18. Aerial view of the Hugo Pegmatite Mine, Keystone, South Dakota.

zones of albite (plagioclase feldspar). This material is crushed and used in the ceramic and glass industries primarily.

Also of commercial importance is "sheet mica" and "scrap mica". The specific mica mineral is muscovite or white mica. "Sheet mica" refers to muscovite which may be cut into sheets of approximately one square inch or larger. Because of its unique physical and electrical properties, sheet mica is particularly useful in the electronic and electrical industries. "Scrap mica" is that which is reduced by grinding to various sizes for use in roofing materials, paint, rubber, and other products.

Although lithium production has been minimal in the Black Hills recently (Norton 1975), total production of lithium minerals over the years has been significant. The primary commercial lithium minerals are spodumene, amblygonite, and lepidolite. Spodumene, the chief commercial source of lithium, is a lithium aluminum silicate, whereas amblygonite is a lithium aluminum phosphate. Lepidolite is a purple mica. Among its wide variety of uses, lithium is of importance in the manufacture of batteries, welding and brazing fluxes, glass, ceramic products, chemicals, and air conditioning systems.

The Black Hills pegmatites have also yielded a considerable amount of the mineral beryl, a major source of beryllium. This metal has been used in various alloys which in turn have been employed in electronic equipment and more recently in aircraft and space equipment. Because of its light weight, rigidity, and electrical, nuclear and thermal properties, beryllium has a great diversity of uses. It is presently used in the nuclear field as well as in many other areas.

Among other minerals mined from the pegmatites is cassiterite, a source of tin. In addition tantalum and niobium, from the mineral tantalite-columbite, have been mined in the region. Other minerals are white quartz and rose quartz, both used in the ceramic industry, and gem varieties of beryl, spodumene, and tourmaline. A currently active mill for processing quartz for ceramics is illustrated in figure 2.19. Many hand specimens of these and other minerals from the granitic pegmatites are readily available in rock and mineral shops which are common, particularly in the southern Black Hills.

Gold Deposits

The Black Hills region has long been noted for its gold deposits. From the onset of mining in 1875 through 1973, South Dakota has produced a recorded 35,484,483 troy ounces, 91% of which was provided by the famous Homestake Mine in Lead, the largest gold mine in North America. This quantity of gold constitutes more than 1% of all the gold mined world-wide throughout history, and defines the Black Hills as a major mining area (Norton 1975). At the current price, this quantity of gold exceeds 12 billion dollars in value. Extraction of silver associated with the gold has amounted to 13,048,000 ounces of which 57% has been a by-product from the Homestake Mine (Norton 1975). Following initial production in 1878, gold production at the Homestake Mine increased steadily for several decades. There was an acceleration of production in the 1930's following an increase in the price of gold to $35 an ounce. Production peaked in 1939 and was followed by a sharp decline during World War II (Norton 1975). By the end of 1950, sharp increases in

Figure 2.19. Dakota Quartz Products Mill, Keystone, South Dakota.

mining costs combined with a static price for gold caused all mining in the Black Hills except in the Homestake Mine to terminate. Production increased generally after 1951 as mining methods were improved, and the overall operation was enlarged at the Homestake Mine. Although production in 1965 exceeded the 1939 peak, it has generally diminished since then. However, the dollar value of the gold produced has increased dramatically because of the substantial price increase.

The major types of commercial gold deposits in the Black Hills include (1) those which occur as primary replacement bodies in ancient Precambrian rocks, and (2) those which occur as modern placer deposits (secondary concentrations in stream gravels formed by erosion and reworking of older primary deposits). Examples of both types of deposits are described later in the guidebook.

Gold was first discovered in the Black Hills in 1874 as placer deposits located three miles east of the present city of Custer (Norton 1975). The much richer placer deposits of Deadwood Gulch were discovered in late 1875. At the very end of 1875, prospectors traced the source of these Deadwood Gulch secondary placers to the primary deposits which are now mined at the Homestake Mine. The original Homestake claim was located in April, 1876 (Norton 1975).

Iron Deposits

Since 1890, when mining began, a relatively small amount of iron ore (on the order of 220,000 tons) has been produced in the Black Hills (Redden 1975). Of this amount roughly 150,000 tons have been used by the State Cement Plant in Rapid City as an additive in the production of cement.

The most important types of Black Hills iron deposits include (1) primary taconite of Precambrian age, and (2) secondary deposits associated with the Cambrian sedimentary rocks (Harrer 1966). Although not yet mined, the taconite deposits, such as those in the Nemo area, have the potential for considerable future development (Redden 1975). The secondary deposits have been mined at scattered localities, such as in the Strawberry Ridge area south of Deadwood, but these have limited potential. Both the Nemo taconite and Strawberry Ridge secondary deposits are described later in the guidebook.

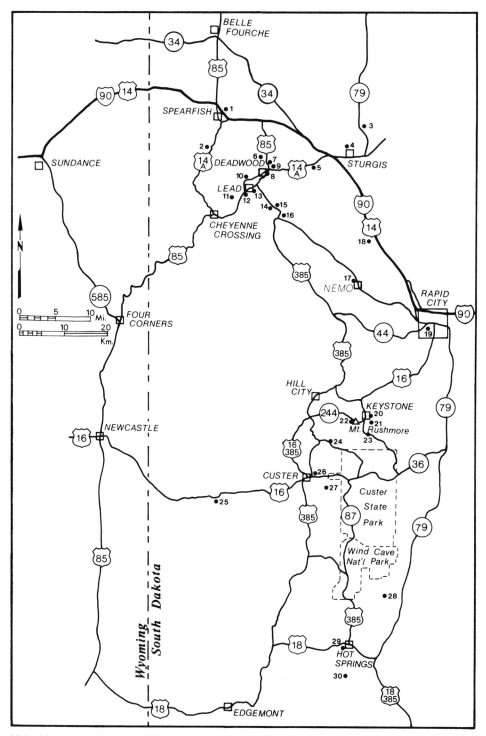

Major highways in the Black Hills region and localities described in the guide.

Chapter **3**

LOCALITY DESCRIPTIONS

Stop 1. Lookout Peak and the early Mesozoic rock sequence, Spearfish

Location. Lookout Peak is the highest point just east of Highway I-90 on the east edge of Spearfish, Lawrence County, South Dakota (fig. 3.1). Access to this locality is by parking in Spearfish, walking across the entrance ramp to I-90, and climbing the hill.

Description. Probably the most obvious landform in the Black Hills is the Red Valley. Lookout Peak provides an unexcelled view of this feature in the northern Black Hills; it offers an excellent opportunity for examination of the rocks that underlie the Red Valley as well as those that form the outer margin of the valley. As one ascends Lookout Peak the following formations are encountered: Spearfish, Gypsum Spring, Sundance, Morrison, and Lakota. The bright red shale of the Spearfish Formation is the rock unit on which the Red Valley is carved. As one looks across the valley toward Spearfish, patches of red material of the Spearfish Formation can be seen exposed throughout the valley. As one ascends the hill, the first ledge encountered is developed in a white, very

soft material—gypsum—marking the base of the Gypsum Spring Formation (fig. 3.2). Although the rock is extremely soft (with a hardness of two) it forms a prominent ledge because the Black Hills region lies in a semi-arid climatic zone in which gypsum is not rapidly dissolved and eroded. The gypsum is highly fractured, and what rainfall does occur percolates rapidly through the material and removes very little of the gypsum. A similar resistant ridge of Gypsum Spring Formation can be observed in many places in the northern Black Hills. The contact between the Gypsum Spring and the overlying Sundance Formation is marked by the change from gypsiferous shale to brown sandy shale at the base of the Sundance Formation. Most of Lookout Peak is composed of the Sundance Formation which extends from the top of the Gypsum Spring Formation to within 100 feet of the top of the hill.

Climbing through the Sundance Formation and examining the rocks reveals considerable variation in rock type. In fact, the Sundance Formation has been subdivided into several members based on this lithologic variation. The primary rock types are tan to red sandstone which is fairly resistant and forms ledges (fig. 3.3) typically,

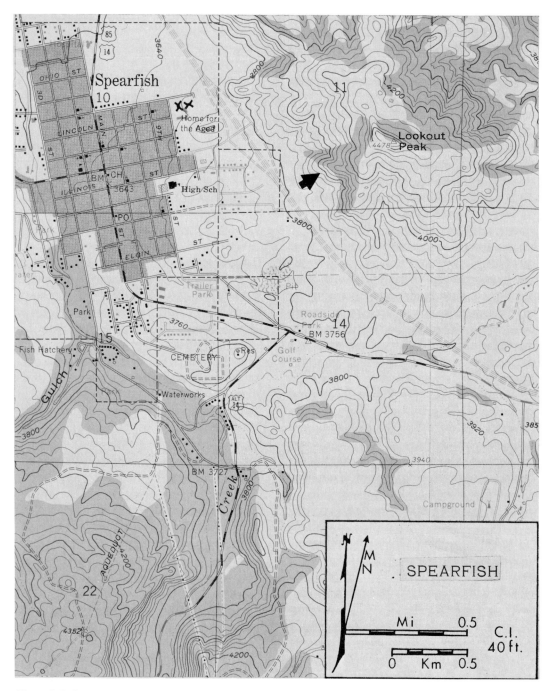

Figure 3.1. Stop 1, Lookout Peak and the early Mesozoic rock sequence.

Figure 3.2. Lower portion of Lookout Peak, Spearfish, South Dakota. In ascending order the rocks exposed are the Spearfish, Gypsum Spring, and Sundance formations.

Figure 3.3. Ledge-forming sandstone in the lower portion of the Sundance Formation. Note the variation of bedding types and thicknesses present.

atus. Elsewhere in the formation star-shaped stem pieces of the crinoid, *Pentacrinus,* and rod-shaped skeletons of the squid-like *Belemnites* (fig. 2.11) can be found. The crinoid stem fragments are rare, but *Belemnites* can be collected locally in enormous numbers. All of the fossils are indicative of a shallow marine or lagoonal habitat as the environment of deposition for the Sundance Formation (Wright 1973).

Above the Sundance, rocks of the Morrison and Lakota form the cap on Lookout Peak. However, the Morrison is extremely poorly exposed, and the rocks of the Lakota form only a small remnant at the top of the Peak.

The view from the top of Lookout Peak offers an opportunity to understand the relationship between several different physiographic features of the Black Hills (fig. 3.4). The Red Valley is viewed clearly as a feature paralleling the rocks in the Black Hills; that is, it follows the pattern of ridges surrounding the Hills. Looking

and finer-grained materials, namely siltstone and shale, which are less resistant to erosion and form gentler slopes. Marine fossils are preserved in the Sundance Formation, although at Lookout Peak only one layer is particularly fossiliferous. At the very top of the formation, a thin layer of calcareous sandstone is dominated by fragments of oysters, *Camptonectes bellistri-*

Figure 3.4. View from Lookout Peak southwest to the Black Hills. The city of Spearfish is located in the Red Valley (foreground) on the dip slope of the Minnekahta Formation. In the background is Crow Peak, a Cenozoic age intrusive body.

across the Red Valley to the main part of the Hills, one sees a broad, gently sloping surface which is the upper surface of the Minnekahta Formation, the youngest of the Paleozoic rocks. The Minnekahta Formation forms this type of sloping surface at nearly all points around the Black Hills, just inside the Red Valley. Similarly, the outer escarpment of the Red Valley can be seen to be the eroded edge of the rock sequence from the upper Spearfish through the Lakota Formation. The outer margin of the Valley, therefore, is defined by an eroded escarpment, and the inner margin of the valley is defined by the dip slope of the uppermost Paleozoic rocks.

From this same vantage point one can also observe that the Paleozoic rocks of the Black Hills are domed into a form representing a turtle back. This general form is interrupted to the westsouthwest by Crow Peak (fig. 3.4), the shape of which does not seem to conform to the normal or predictable form of the Black Hills uplift. Crow Peak is one of a number of Tertiary intrusive bodies which locally dome the overlying sedimentary rocks.

Stop 2. Bridal Veil Falls, Spearfish Canyon

Location. Access to this locality is via U.S. Highway 14A (Spearfish Canyon Road). From the junction of U.S. 14 and 14A southeast of Spearfish, travel south on 14A 5.9 miles to Bridal Veil Falls (fig. 3.5).

Description. The trip up Spearfish Canyon is perhaps one of the most scenic in the Black Hills region. As one enters the

canyon the rocks change abruptly from the red sediments of the Spearfish Formation underlying the Red Valley, to the thin limestone unit, the Minnekahta Formation, then to the Opeche and Minnelusa formations, and finally to the Pahasapa Formation. In the lower reaches of the canyon this section is traversed rather rapidly because the beds dip steeply to the north but as one gets farther into the canyon the beds appear to level out, and most of the drive to Bridal Veil Falls traverses rocks of Mississippian age, the Pahasapa Formation. As one approaches the area of the Falls, however, older rocks of the Deadwood Formation appear at road level. At Bridal Veil Falls sediments of the Deadwood Formation can be examined at road level on the west side of the road (fig. 3.6). They contain large quantities of a green mineral, glauconite, which gives freshly broken rock samples a dark green color and which produces the reddish and brown colors of the rock upon weathering. The Deadwood at this locality contains abundant tracks, trails, and burrows of trilobites and other Cambrian organisms (fig. 3.7).

Directly east of this exposure, Bridal Veil Falls can be observed (fig. 3.8). Most streams tend to enter other streams at grade, that is, at the level of the master stream. The presence of waterfalls therefore, indicates that some unusual geologic condition exists. In the case of Bridal Veil Falls, this condition exists because the tributary stream enters Spearfish Creek at the location of an intrusive igneous body. This igneous rock, phonolite, is much more resistant to erosion than the surrounding Deadwood Formation, and the stream, which carries water only intermittently,

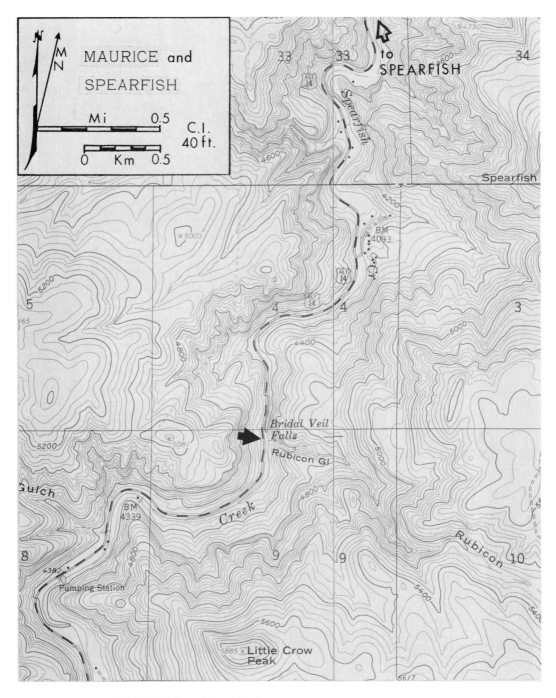

Figure 3.5. Stop 2, Bridal Veil Falls and Spearfish Canyon.

Figure 3.6. The Deadwood Formation exposed directly across the highway from Bridal Veil Falls. These rocks contrast markedly with those beneath the falls.

Figure 3.7. Trails and burrows exposed on the underside of a loose block of Deadwood sandstone. Note that several different types of traces are present.

Figure 3.8. Bridal Veil Falls. Note that the rocks on which the falls is formed are vertically jointed but show no layering as the Deadwood rocks above the falls and in figure 3.6.

has not cut down through the intrusion at the same rate that Spearfish Canyon has been cut. The result is Bridal Veil Falls.

The igneous body can be identified by its color, which differs from that of the surrounding Deadwood Formation, and by the fact that it lacks the distinctive sedimentary layering evident in the Deadwood Formation. Beds of the Deadwood Formation appear to be almost horizontal above and to the north of the intrusion, whereas the only breaks that can be observed in the igneous body are nearly vertical joints. This igneous body is one of numerous small Cenozoic intrusions which were emplaced in the northern Black Hills during the Eocene Epoch. Some were extensive

enough to produce large structures such as Elkhorn Peak near Whitewood and Bear Butte northeast of Sturgis, whereas most were smaller bodies that affected only very limited areas such as the one seen at Bridal Veil Falls.

The area of Bridal Veil Falls is also an excellent place at which to observe the effects of landsliding or gravity movements. From this vantage point and by traveling north and south along 14A, numerous fresh landslide scars can be seen, primarily on the west side of the valley. These landslides, all of generally small scale, can be identified by the presence of steep slopes, generally light-colored and stripped of vegetation, with associated rock debris consisting of large, angular blocks at the base of the slope and in the stream channel (figs. 3.9 and 3.10). Although the landslides are not extremely large, they do have a significant effect on man's use of the area. Whenever a landslide occurs, considerable damage is done to the highway, and the stream may be temporarily dammed with debris. This necessitates clearing the stream channel and reconstructing the highway. Many such reconstructed areas are marked by the presence of fresh patches of asphalt along the highway. Many of these landslides occur in late spring and early summer when melting snow saturates the rock on the steep walls of the canyon. This rock may have been broken loose earlier in the season by freezing and thawing. This time of the year also coincides with the heaviest rainfall and most active flooding. Evidence of flooding is abundant throughout the valley. At the entrance to the valley, for example, a bridge, now destroyed (fig. 3.11), has been

Figure 3.9. Small landslide near Bridal Veil Falls. The area of the slide is defined by the lighter-colored rock and the absence of vegetation (arrow).

Figure 3.10. Small cone of debris at the base of a landslide in Spearfish Canyon.

Figure 3.11. Bridge across Spearfish Creek near the mouth of Spearfish Canyon. Spring flooding destroyed the structure which has now been bypassed.

simply bypassed by construction of a permanent "detour." Attempts have been made to reduce the effects of this stream erosion, and they are moderately successful. Chief among these is stabilization of the stream valley by fastening stout wire mesh over stream bounders to lock the material together and render it less mobile. Ultimately, however, the grating is destroyed or undercut and stream erosion again adversely affects the highway.

Stop 3.   Bear Butte

Location.   Bear Butte State Park, South Dakota, on S. Dak. Highway 79 about seven miles northeast of Sturgis, Meade County, South Dakota (fig. 3.12). The geology of this area can be observed conveniently from two vantage points. Traveling along the west margin of the Butte on South Dakota 79 provides a fine view of the Butte and some of the flanking structures. By far, however, the best view of the Butte can be obtained by taking the hiking trail to the top of the Butte. Access to the hiking trail is gained by checking in at the park headquarters and following the marked trail signs.

Description.   Bear Butte (fig. 3.13) is one of the numerous intrusive igneous bodies injected during early Cenozoic time, probably the Eocene Epoch, in the Black Hills region. Bear Butte is the easternmost of a series of related intrusive bodies that form a narrow band (10 miles wide) extending westnorthwest across the northern Black Hills. The rock of which Bear Butte is composed has been classified as rhyolite (Darton and Paige 1925, p. 19) (fig. 3.14), the

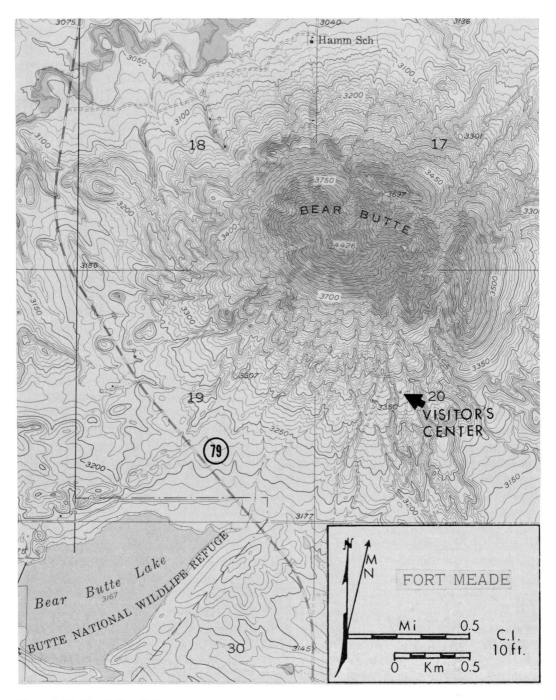

Figure 3.12. Stop 3. Bear Butte.

Figure 3.13. Bear Butte viewed from the west. The top of the butte is about 1200 feet above the point from which the photo was taken. The rocks in the foreground are Mesozoic sedimentary rocks, primarily, that have been folded between the Butte and Bear Butte Circus.

Figure 3.14. Fine-grained rhyolitic rocks forming the core of Bear Butte. Note the prominent joint, or fracture, pattern.

fine-grained equivalent of granite. Only locally is the rock porphyritic; thus confirmation of the rock type must be made microscopically. The Bear Butte intrusion apparently has the form of a cylindrical mass which was forcefully injected or pushed up through the surrounding sedimentary rocks (Darton and Paige 1925). Although the Butte presently rises about 1200 feet above the surrounding landscape, it probably pushed the sedimentary rock upward several thousand feet. As the material was intruded it was not particularly hot by comparison with most igneous intrusions, because its contact with the surrounding sedimentary rocks shows only minor metamorphic alteration of these rocks. In some places baking and addition of siliceous cement to the sediments indicates some effect of the intrusion, but these are relatively minor features which extend no more than a few tens of feet beyond the margins of the body.

The intrusion was forcefully injected as indicated by the fact that rocks ranging in age from Mississippian (Pahasapa Limestone) to Cretaceous (Pierre Shale) abut structures associated with the intrusion. Apparently, as the mass forced its way upward through the sedimentary rock sequence, it elevated and extended the overlying sedimentary rocks into a highly complicated structure. On the east side of the Butte, clearly visible from the hiking trail, is an exposure of the Pahasapa Limestone oriented nearly vertically against the side of the intrusion (fig. 3.15). This is surrounded by progressively younger rock units to the east; but to the south the Pahasapa is truncated abruptly against Cretaceous shale units, the Niobrara and Carlile formations.

Figure 3.15. Large mass of Pahasapa Limestone (arrow), dipping nearly vertically into the ground, on the east side of Bear Butte.

This style of structural deformation is quite different from that viewed from the top of the Butte looking to the west. In this region one can observe a large concentric structure, Bear Butte Circus (fig. 3.16), which apparently reflects the presence of another plug of igneous material similar to Bear Butte but still buried at some depth below the surface. This uplift, which probably accompanied emplacement of the Bear Butte intrusion, domed up the sedimentary rocks above it about 1000 feet. The outline of Bear Butte Circus is well defined by the relatively resistant sandstone units in the Sundance, Lakota, and Fall River formations.

The relationship of Bear Butte and Bear Butte Circus to the surrounding sedimentary rocks clearly demonstrates that this intrusive event was quite independent of the processes of uplift of the Black Hills themselves. The same is true of other small-er intrusive bodies within the central Black Hills. This conclusion follows from the fact that the local Bear Butte structures appear to be superimposed on the general Black Hills structures; that is, the normal structural setting of rocks as one moves away from the uplift are consistent with those associated with the Black Hills Uplift but change abruptly in the vicinity of Bear Butte and are related directly to the intrusive event.

If, instead of climbing to the top of Bear Butte one proceeds north on South Dakota 79 and around its west edge, one traverses the area of Bear Butte Circus. Although access to most of the exposures is on private land and cannot be accomplished without permission, from the road it is possible to observe the attitude of the beds rimming the Circus flats and to determine that, south of Bear Butte, the beds dip south whereas north of the Butte the beds

Figure 3.16. Panoramic view from the west side of Bear Butte with the Black Hills in the background and Bear Butte Circus (arrows) in the foreground.

dip north. The same rocks are encountered north and south of the Butte confirming the domal structure of Bear Butte Circus.

Stop 4. Sly Hill Overlook, Sturgis

Location. North edge of Sturgis, Meade County, South Dakota. Access to this locality is via the old Bear Butte road that originates in Sturgis as the northward continuation of Junction Street (fig. 3.17).

Description. The road cut on the climb up Sly Hill exposes, in order, the Morrison Formation, the Lakota Formation, and the Fall River Formation. The Morrison is poorly exposed although small patches of it can be seen through the vegetation just below the massive sands of the Lakota Formation. Sly Hill is part of the outer rim of the Black Hills commonly referred to as the Dakota Hogback, or Escarpment. Typically, the rocks that support the escarpment are the Lakota and Fall River sandstones.

The Lakota Formation on Sly Hill consists of two distinctly different rock types. The lower portion of the unit consists of sandstone with some interbedded conglomerate which forms either massive layers as much as three feet thick or cross-bedded layers (fig. 3.18). Examination of the surface of the Lakota exposure suggests that it is very well cemented sandstone because the surface is quite hard and bits of rock material can be broken away only with difficulty. Upon doing so, however, it is clear that the rock is relatively friable, that is, it crumbles readily, and that the surface has been secondarily hardened by a process referred to as "case-hardening." Case-hard-

ening is a result of movement of calcium-rich water toward the surface of the exposure, evaporation of the water, and precipitation of calcium carbonate as a surficial cement. Above the sandstone of the lower Lakota Formation is a sequence of interbedded sandstone, siltstone, shale, and lignitic shale which comprises the Fuson Member of the Lakota Formation (fig. 3.19). The contact betwen the two units is not visible on Sly Hill. Above the Fuson, the Fall River Sandstone consists of more thinly bedded and cross-bedded reddish-brown sandstone (fig. 3.20). The Fall River can be distinguished from the Lakota both because of its position relative to the Fuson Shale and its distinctly different bedding characteristic.

These formations, the Lakota and the Fall River, represent the earliest Cretaceous rocks in the Black Hills region. They were deposited as fluvial, near shore, sand bodies on the margin of the advancing Cretaceous seaway. Prior to the time of their deposition, the area had been elevated somewhat above sea level, and sediments of the Morrison Formation had accumulated on a surface of generally low relief. In early Cretaceous time, the Black Hills region was once again subjected to advancing marine conditions which persisted through much of the Cretaceous until the Laramide Orogeny uplifted the area above sea level for the last time.

Sly Hill also provides an excellent overview of the geology of the northeast corner of the Black Hills uplift. Standing on this part of the Dakota Hogback one can look south and see the continuation of the hogback marking the outer rim of the Red Valley, the Red Valley developed on the Triassic Spearfish Formation, the dip slope

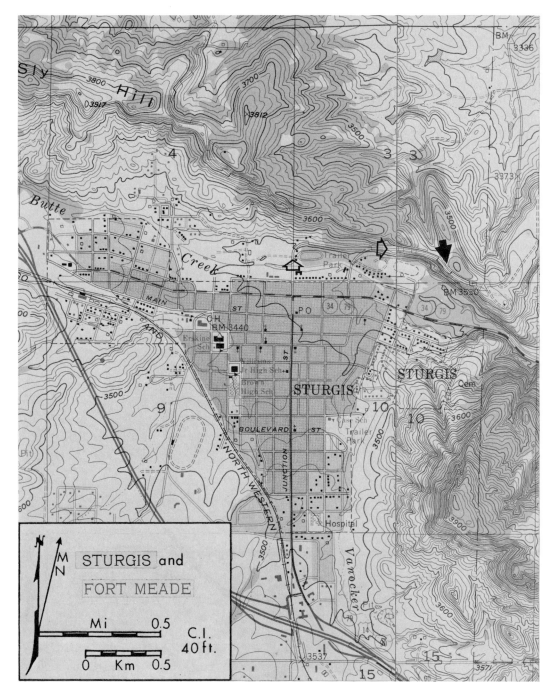

Figure 3.17. Stop 4, Sly Hill overlook, Sturgis, South Dakota.

Figure 3.18. Lakota Sandstone exposed along road on Sly Hill, Sturgis. This rock is the dominant ridge-forming rock of the escarpment rimming the Black Hills.

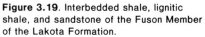

Figure 3.19. Interbedded shale, lignitic shale, and sandstone of the Fuson Member of the Lakota Formation.

Figure 3.20. Fall River Formation. Contrast the style of weathering of the Fall River with that of the Lakota Formation in figure 3.18.

or upper surface of the Minnekahta Formation, and the lower reaches of the valley through which Bear Butte Creek flows. If one continues to the summit of Sly Hill and along the road to the north for about two miles, the terrain changes abruptly as do the exposed rock units. Although the beds continue to dip east in this area, one has traveled beyond the obvious limits of the Black Hills Uplift to a region that would perhaps more appropriately be referred to as the Great Plains. The rocks underlying this area dip very gently and are dominantly blue-gray shale of the Graneros Formation. This marine shale is reminiscent of many of the Cretaceous shale units exposed from the flanks of the Black Hills eastward to the Missouri River and, for that matter, exposed on all other sides of the Black Hills Uplift. Examination of the Graneros

Formation indicates that it is abundantly fossiliferous, containing disarticulated bones and scales of fish along with very occasional impressions of mollusks and arthropods. From this vantage point an excellent view of Bear Butte is also available.

Stop 5. Late Paleozoic Rocks and the Boulder Park Syncline

Location. Boulder Canyon and Boulder Park along Bear Butte Creek and U.S. Highway 14 about three miles west from Sturgis, Lawrence County, South Dakota (fig. 3.21).

Description. The geologic section exposed along U.S. Highway 14 through this area

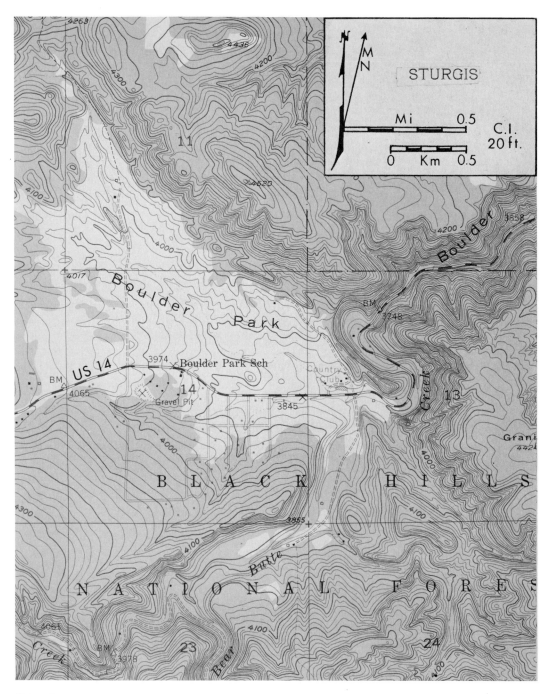

Figure 3.21. Stop 5, Boulder Park.

Figure 3.22. Panoramic view of Rainbow Arch. The rocks shown in the view are those of the Minnelusa Formation.

includes rocks of the Pahasapa Limestone (Mississippian age), Minnelusa Formation (Pennsylvanian), Opeche Formation (Permian), Minnekahta Formation (Permian), and Spearfish Formation (Triassic). The Minnelusa Formation here consists of about 450 feet of red, white, and brown sandstone and shale which tends to form steep but deeply weathered cliffs (fig. 3.22). The lower portion consists of sandstone and shale interbedded with limestone very much reminiscent of that which forms the upper part of the underlying Pahasapa Limestone. In fact, the contact between these two units is interfingered such that Pahasapa lithology alternates with Minnelusa lithology through about 20 feet of the lower Minnelusa. Above this a prominent red silty shale is clearly exposed in most thick sections of the formation. Overlying that, the unit is predominantly a dull red to white sandstone which weathers as thin layers of differing resistance.

The overlying Opeche Formation is conformable with the Minnelusa but separated from it by a distinct break. It consists of about 100 feet of fissile calcareous shale

and siltstone (Unit 1, fig. 3.23). The unit weathers rapidly and is typically expressed as a valley between the Minnelusa Formation and the overlying Minnekahta Formation. Exposures of this formation are, however, easy to identify because of their bright red color and because of the position of the rocks between the Minnekahta and the Minnelusa.

The youngest Paleozoic rocks in the Black Hills are included in the Minnekahta Formation (Unit 2, fig. 3.23), a relatively thin unit, typically 30 to 40 feet thick, but one that is extremely important in terms of the landforms developed around the margin of the Hills. As one travels through the Racetrack, for example, the narrow ridges rising around the margins of the Hills are produced on the Minnekahta Formation. Close examination of this unit indicates that it is composed of limestone containing several interesting features which mark it as of shallow-water marine origin. Typically, the lower part of the unit contains massive, extremely fine-grained limestone overlain by very thinly laminated limestone (fig. 3.24) which appears to have been

Figure 3.23. Boulder Park syncline. The rocks exposed along the road, U.S. Highway 14, at the lower right are the Opeche Formation (1) and the Minnekahta Formation (2).

Figure 3.24. Thinly laminated Minnekahta Formation along U.S. Highway 14, north of Deadwood.

Figure 3.25. Algal mound in the Minnekahta Formation along U.S. Highway 14 north of Deadwood.

formed as algal mats, i.e., thin sheets of material deposited over the sea floor as a kind of limey scum due to the action of calcareous algae. Above this these same algal mats form dome-shaped mounds (fig. 3.25) with a diameter of about six or seven inches. Above these, algal mounds and mats have been broken, fragmented, and redeposited in jumbled irregularity. This type of sequence is not uncommon in carbonate rocks deposited in shallow-water marine environments.

One other feature of the Minnekahta that is obvious in many exposures, notably in the Boulder Park region, is that the thin to flaggy beds commonly are distorted into tight folds. These folds have apparently developed as minor structures related to the main Black Hills Uplift.

Perhaps the most interesting and obvious geologic feature in the Boulder Park region is a north-south trending anticline referred to locally as Rainbow Arch (fig. 3.22) and the adjacent Boulder Park Syncline (fig. 3.23). Rainbow Arch is the western limb of a broad anticlinal structure beginning at the margin of the Black Hills Uplift near Sturgis. In that area the Minnekahta Formation, which can be used to define the structure, dips to the northeast at an angle of about 10° to 15°. As one approaches the Boulder Park area, about three miles southwest, the Minnekahta and the underlying Minnelusa formations dip into the ground at an angle of about 55° southwest. Continuing into the area of the Boulder Park Country Club, the highway crosses the axis of Boulder Park Syncline,

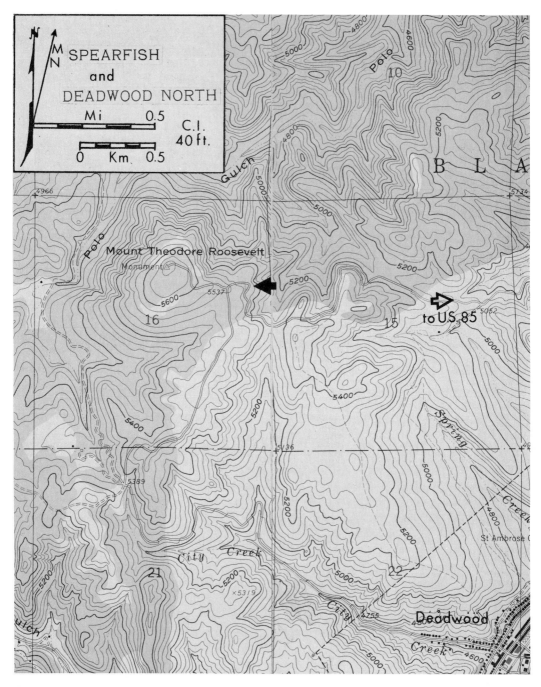

Figure 3.26. Stop 6, Sheep Mountain Stock on Mount Theodore Roosevelt.

again defined by Minnekahta ridges which extend from the Country Club both northwest and southwest. Dips in this area are typically gentle but inclined both north and south toward the highway. The effect of this folding is that the rock sequence, normally seen only around the margin of the Black Hills Uplift, is repeated as one travels from Sturgis to Deadwood on U.S. 14.

Stop 6.   Sheep Mountain Stock

Location. The Sheep Mountain Stock is well-exposed on Mount Theodore Roosevelt. It may be approached by means of an unpaved road (marked by a sign indicating Mount Theodore Roosevelt) north of Deadwood from U.S. Highway 85, or reached from the downtown area of Deadwood by following a road along City Creek (fig. 3.26). The specific locality described is near the top of the peak—just east of the intersection of the main unpaved road and a spur which leads directly to the monument at the top of the peak. Following the first access route, this locality is 1.6 miles west of Highway 85 (fig. 3.26).

Description. The Sheep Mountain Stock (fig. 3.27) and the Cutting Stock (Stop 10) constitute two major stocks in the Lead area. The Sheep Mountain Stock is a composite igneous body composed of two main rock types intruded at different times within a single Cenozoic igneous cycle. The first of the two rock types is well-displayed in a road cut at the locality marked on figure 3.26. Here the rock is extremely fine-grained and lacks minerals visible with the unaided eye. Microscopic examination shows it to be rhyolite dominated by quartz

and feldspar. As is typical of these Cenozoic intrusions, the rock here is highly fractured and weathers to angular rubble.

One can observe the second rock type 0.4 miles further west toward the road to the monument. Here the rock is a distinctive rhyolite porphyry characterized by orthoclase feldspar and dark, smokey quartz phenocrysts which exhibit well-developed crystal terminations. The quartz phenocrysts, or "eyes," are up to 1/4 inch in diameter (fig. 3.28). The age relationship between the two rock types is established by the fact that, locally, dikes of the rhyolite porphyry cut across the non-porphyritic rhyolite, indicating that the former was intruded after the latter had been intruded and solidified.

Stop 7.   Whitewood Creek Gold Deposit

Location. Access to this locality may be obtained by turning off U.S. Highway 14A approximately one mile northeast of Deadwood. At this point, marked by a historic plaque opposite a cluster of gasoline stations, turn north and travel along a dirt road which earlier served as an embankment for the Chicago and Northwestern Railroad (fig. 3.29). The site is located at a broad, open meander of Whitewood Creek slightly more than one mile from Highway 14A. Permission to enter the area must be obtained from Mr. Morris Hoffman, manager and owner of the Strawberry Hill Mining Company which operates the gold-processing equipment situated on the stream bed here.

Description. This occurrence is part of a larger deposit which constituted the initial discovery of gold in the northern Black

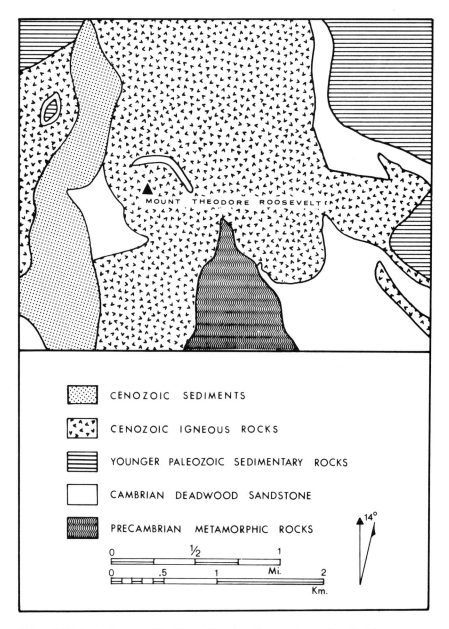

Figure 3.27. Geologic map of the Mount Theodore Roosevelt area (modified from Homestake Mining Company unpublished map).

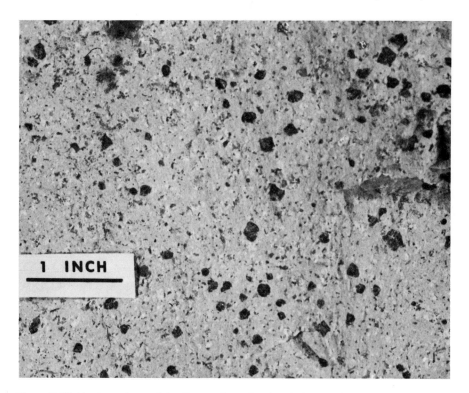

Figure 3.28. Hand specimen of rhyolite porphyry showing distinctive phenocrysts (or "eyes") of smokey quartz.

Hills. Although the source of this secondary stream placer deposit is the basal gold-bearing conglomerate of the Cambrian Deadwood Formation, further exploration of these placer deposits ultimately lead to discovery of the older Homestake ores mined today in Lead (Noble 1950).

At this locality the gold occurs as clastic grains which are concentrated primarily within gravel beds and lenses ranging typically from two feet to 15-20 feet in thickness. A large amount of gravel has been localized here because, further downstream, a landslide dammed the stream. The gold ranges in grain size from minute fragments up to small nuggets of roughly one-quarter inch diameter (fig. 3.30). As is typical of placer minerals, the gold occurs as small flakes and flattened masses due to pounding action during stream transport. Some of the gold here is actually an amalgam, or a mixture of native gold with mercury. The mercury was derived from milling of ores upstream in the early days.

The processing begins by shoveling the gold-bearing gravel using a front-end loader. The material is then dumped into a funnel-shaped device which feeds it to a shaker bin and ultimately onto a conveyor

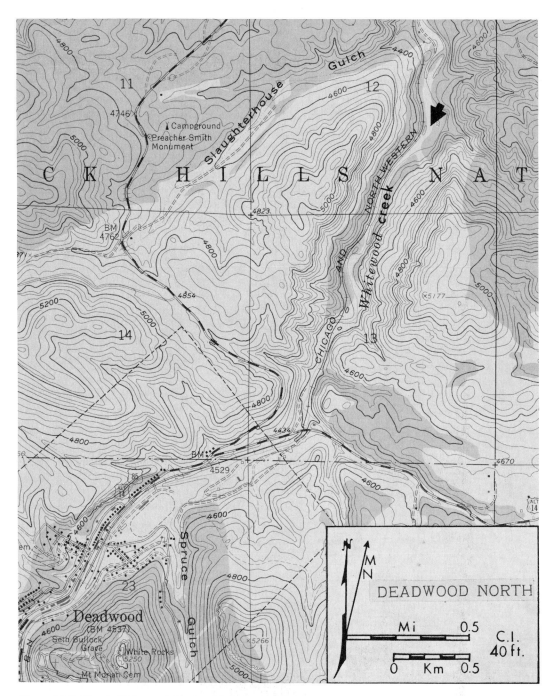

Figure 3.29. Stop 7, Whitewood Creek gold placer.

Figure 3.30. Gold flakes and nuggets, about one-fourth inch in diameter, extracted from gravels in the Whitewood Creek gold placer.

belt. The conveyor transfers it to a trommel, an inclined, revolving, cylindrical screen, which allows all material less than one-half inch in diameter to fall through it (fig. 3.31). As water is added to the gravel in the trommel, the large cobbles and boulders are delivered at the far end and returned to the stream bed by conveyors. The fine-grained gold, sand, finer gravel, and water fall through the holes in the trommel and are fed onto a table covered with astroturf and divided by riffles. The heavier gold is trapped in back of the riffles while the water and lighter-weight quartz grains move downslope off the table. Ultimately, the riffles are removed and the astroturf, which has trapped the fine gold in back of the riffles, is emptied into a container. This initial gold concentrate is then taken to the mill where it is further processed by use of an automatic panner. The final product is then assayed (to determine the percent of gold present), melted, poured into bars, and taken to the Homestake Mine for sale. A small amount of this gold is also sold to mineral collectors and to others for jewelry making.

Under a variety of ownerships, this placer deposit has been mined sporadically since the early 1950's. Equipment related to earlier, unprofitable mining ventures lies strewn along the stream bed in this area.

Figure 3.31. Placer mining equipment showing the intake conveyer (at the far end), trommel, and waste-rock conveyer for removal of coarse material (in the foreground).

The present company, which has rights to a stretch of these gravels slightly more than two miles in length, has cleaned up the area considerably and is beginning to generate a return on its investment. At present production amounts to an average of ten ounces of gold per day.

Stop 8.   The Deadwood Formation

Location. Northeast end of Deadwood, South Dakota. Access to the geologic section is via Main Street (U.S. Highway 85-14A) in Deadwood just southwest of the rodeo grounds (fig. 3.32). The lower part of the section is well exposed on the south side of an unpaved road paralleling Whitewood Creek south of Dunlop Avenue (locality 8A). The upper part of the section is located directly across the highway on the high bluff overlooking the city (locality 8B; fig. 3.33).

Description. The Deadwood Formation includes the oldest sedimentary rocks of the Paleozoic section in the Black Hills. It consists of about 400 feet of sandstone, dolostone, dolomitic sandstone, and shale at this locality but may vary from five to 400 feet throughout the Black Hills. The sequence of rocks is such that these various lithologies are interbedded with one another and produce a complex section with

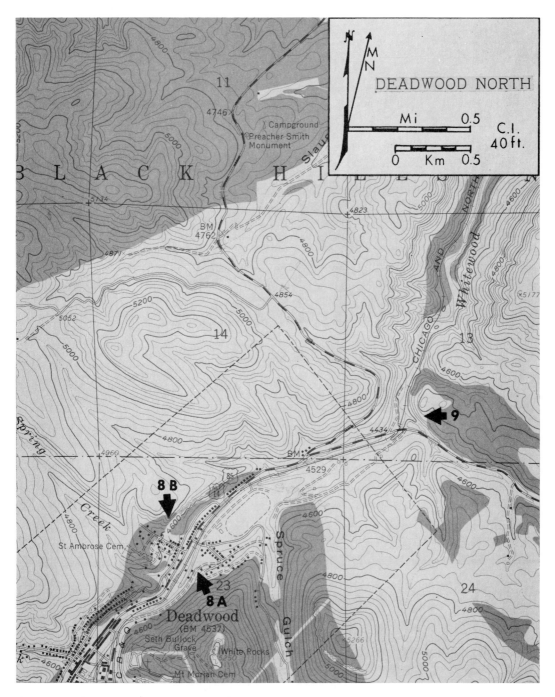

Figure 3.32. Stops 8 and 9, the lower Paleozoic rocks of the Deadwood area.

Figure 3.33. The Deadwood and Alladin formations in Deadwood. Note the differences in resistance to weathering of layers within the Deadwood.

marked lithologic variation occurring in a matter of a few inches as one progresses up the section (fig. 3.34). Examination of the rocks, however, suggests that they were all formed in a relatively shallow water environment in which sediment was deposited as a series of cross-bedded layers of sandstone interbedded with units that are somewhat finer grained and flat bedded. Fossils are not common in the unit, but near the base of the section fragments and an occasional entire specimen of tiny brachiopods have been collected. Throughout the section occasional fragments of trilobites can also be found. The major evidence of organic remains in the unit consists of tracks, trails, and burrows apparently produced by organisms working their way through the sediment at the time it was being deposited to extract food. Their traces remain as

filled, irregular-shaped burrows and tunnels (fig. 3.35).

Although the Deadwood Formation is thickest in the northern Hills and tends to become somewhat thinner to the south, it is found throughout the Black Hills in contact with the ancient Precambrian rocks. (fig. 3.36). Wherever seen, it has the same distinctive reddish-brown weathering color that is evident in the Deadwood area and can be readily distinguished from the rocks above and below.

Directly overlying the Deadwood Formation is a very thin unit, typically about 12-15 feet thick, called the Alladin Formation (fig. 3.37). The Alladin contrasts markedly with the Deadwood Formation in two ways. The Deadwood Formation consists dominantly of sandstone composed of quartz, glauconite (a distinctive green

Figure 3.34. Deadwood Formation near the top of the unit. Differences in resistance to weathering reflect differing degrees to which the rocks have been cemented.

Figure 3.35. Traces of organisms preserved in the Deadwood Formation. These types of remains are structures probably produced as organisms scoured the sediment in search of food.

Figure 3.36. Contact between the Deadwood Formation and the Grizzly Formation of Precambrian age. Note the large quartz boulders that mark the base of the Deadwood. Arrows define the position of the contact.

Figure 3.37. *Skolithos* borings in the Alladin Formation.

mineral), and varying amounts of clay, dolomite, and other minerals. By contrast, the Alladin Formation is almost entirely formed of quartz grains, and the rock is cemented by quartz, making it an exceedingly durable rock unit. The second, and perhaps most obvious, difference between the Alladin Formation and all other viewed in the Black Hills is that it is riddled with tiny burrows (diameter of 1-2 mm) which tend to be closely set and oriented perpendicular to the bedding planes in the rock (fig. 3.37). These structures have been interpreted as the remains of a burrowing organism similar to tube-worms which live in intertidal regions around the oceans of the world today. Therefore, the Alladin Formation is readily distinguishable from the rock below it and, as can be demonstrated elsewhere, is also distinct from the rocks above.

The Deadwood Formation is particularly significant because it records events surrounding the first major marine flooding of the North American continent in Paleozoic time. The sedimentary rocks of the Deadwood Formation, and units very similar to the Deadwood, can be found in many parts of the western interior, including the margin of the Bighorn Mountains and along the Rocky Mountain front. As the North American continent was progressively flooded during Cambrian time, sand was deposited over the area. The sand was probably derived from a combination of weathered material developed on existing Precambrian rocks as well as material eroded from other parts of the continent and deposited in this interior seaway by rivers and streams eroding the upland surfaces. The fluctuations in bedding and rock type seen in the Deadwood Formation are suggestive of unstable or variable

conditions that must have prevailed during Cambrian time.

Fossils collected from the unit are much like those collected from Upper Cambrian rocks in other parts of the midcontinent. Therefore, they permit us both to date the rocks and to establish correlations between these rocks and rocks of similar age found in other parts of the area.

Stop 9. Winnipeg, Whitewood, Englewood, and Pahasapa Formations

Location. About one-half mile northeast of Deadwood, South Dakota, on U.S. Highway 14. Access to this locality is available by parking on the broad parking area south of U.S. Highway 14 just beyond the bridge over the railroad tracks and Whitewood Creek (fig. 3.32, Stop 8). North of the highway, a complete section of the above rocks is exposed from creek level to the top of the cliff. Alternatively, the same sequence of rock, less well exposed, can be observed along U.S. Highway 85 from the north end of Deadwood at the junction with U.S. Highway 14 to the overlook about one-half mile northeast of the junction.

Description. The rocks from the Winnipeg Formation through the Pahasapa Formation, as exposed in this area, represent as complete a sequence of these rocks as can be found anywhere in the Black Hills. In other areas some of the units, notably the Winnipeg and Whitewood formations, are much thinner or absent. The Englewood Formation is also reduced in thickness (or is absent) as one travels into other parts of the Hills. However, in the northern Hills most of these rocks are preserved and

Figure 3.38. The Winnipeg Formation (below) and Whitewood Formation (above). Note the marked difference in weathering of the Winnipeg Shale and the Whitewood Dolostone.

exposed for study. The Winnipeg Formation, at the base of this sequence (fig. 3.38), is a shale unit. It is typically a greenish, plastic, clayey shale whereas the shale above tends to be silty (more abrasive-feeling) and dark brown in color. Although large fossils are absent or exceedingly rare in the Winnipeg, microfossils have been collected from it and establish it as Ordovician age. Microfossils consist predominantly of tiny tooth-like remains of a group of organisms referred to as conodonts. Although conodonts have never been convincingly related to any known group of organisms, they are splendid index fossils; i.e., they tend to be quite diverse in form, and of the many forms of

conodonts, each tends to range through a relatively short interval of geologic time. Thus they can be used readily to determine the age of a geologic sequence.

Overlying the Winnipeg Formation, the Whitewood Formation is strikingly different (fig. 3.38). It consists of about 50 feet of thin-to-massive dolostone and sandy dolostone which weathers to a creamy yellow color. The rocks of the Whitewood Formation are locally riddled with tubes and burrows of organisms that fed on organic debris in the sediments. Although identification of the individual types of organisms responsible for producing the burrows is difficult, it is certainly possible to pick out three or four different types of trails and

Figure 3.39. Large straight cephalopod from the Whitewood Formation.

burrows, suggesting a rather diverse assemblage of organisms as the tracemakers. In addition to these forms, several other fossils have been identified and can be collected from the unit. They include large straight cephalopods (fig. 3.39), typically at least two varieties of gastropods (figs. 3.40 and 3.41), a large form *Maclurites* up to one foot in diameter, and a smaller high-spired form *Murchisonia,* the coral *Halysites,* and the calcareous alga *Receptaculites* (fig. 3.42). This suite of organisms has been found in Ordovician rocks of many parts of the midcontinent and firmly establishes the age of the Whitewood Formation as Middle Ordovician.

Above the Whitewood Formation a sequence of dark-gray shale, approximately 25 feet thick, overlain by reddish limestone also about 25 feet thick, constitutes the Englewood Formation. The contact between the Whitewood and Englewood formations appears to be perfectly parallel to the bedding of the rock units, and no significant relief can be observed on this contact. However, collection of fossils in the Englewood Formation reveals that it is Devonian to Mississippian in age and that the contact between the two units represents a major break in the geologic record— a record spanning the entire Silurian and part of the Devonian periods. This kind of contact, referred to as a paraconformity, is demonstrable only on the basis of fossil content of the rocks above and below the break.

As one examines the upper portion of the Englewood Limestone, it can be ob-

Figure 3.40. *Maclurites*, a large gastropod common to the Whitewood Formation.

Figure 3.41. *Murchisonia*, a high spired gastropod from the Whitewood Formation.

Figure 3.42. *Receptaculites*, the "sunflower coral" which is, in reality, a calcareous alga from the Whitewood Formation.

served that the unit grades into a lighter-colored limestone containing abundant, tiny fragments of fossil organisms. Close examination of these remains indicates that most of them are broken pieces of echinoderms, notably crinoid stem fragments. This change in rock type marks the beginning of the Pahasapa Formation, also of Mississippian age. At first observation, the 200 feet of Pahasapa Formation exposed on this bluff appear to be more or less uniform (fig. 3.43). If one examines the rocks carefully, however, it is possible to observe many changes in rock type throughout this section. For example, the fossil fragments at the base of the formation give way farther up in the section to a uniformly fine-grained limestone and then to a limestone with very thin but distinct laminae, probably produced by growth of calcareous algae. In the upper part of the formation layering is less continuous; occasionally one can find mounds of coralline material which appear to be small reef mounds dominated by the colonial coral *Syringopora*. Throughout the Pahapsa Formation one can also observe small solution features, or vugs, in which beautifully formed calcite crystals have been precipitated. These vugs are due to solution of the rock and reprecipitation by groundwater activity of some of the calcium carbonate. This same proc-

Figure 3.43. The Pahasapa Limestone.

ess, on a much larger scale, has produced the many caves around the margin of the Black Hills.

Stop 10.   Cutting Stock

Location.   Access to this locality is by means of an unpaved road approximately 0.5 mile southwest of Central City off U.S. Highway 14A (fig. 3.44). This road is marked by a sign to the Northern Hills Sanitation Transfer Station. Follow the road along Deadwood Creek bearing left but staying on the right (north) side of the stream until reaching an open area where the road crosses the stream. This location is approximately 0.8 mile from U.S. Highway 14A.

Description.   The rock at this locality is part of a body referred to as the Cutting Stock, a complex of sills, dikes, and irregular masses intruded during Cenozoic time and exposed across an area of roughly eight square miles. The Cutting Stock lies near the center of a major dome that has dimensions 10 miles by 12 miles (Slaughter 1968). At this locality the freshest rock is exposed just east of the stream where the road crosses it (fig. 3.45), in an area that was formerly a talus slope. The rubble on the slope has been removed and used as road fill in the Central City area. The fresh rock here is a medium-grained, quartz monzonite which contains, primarily, zoned plagioclase and orthoclase, quartz, biotite, and hornblende (fig. 3.46). The rock has a moderately large number of small, dark

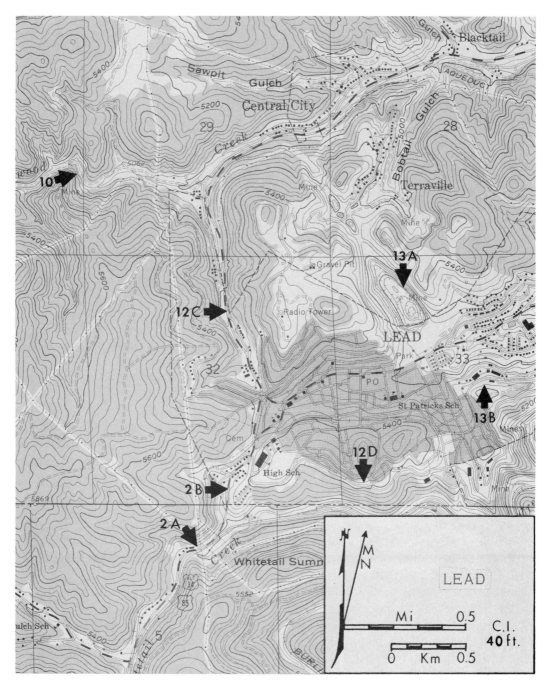

Figure 3.44. Stop 10, the Cutting Stock, west of Central City, South Dakota.

Figure 3.45. Fresh exposure of well-jointed quartz monzonite of the Cutting Stock.

inclusions of hornblende-rich rock. Inclusions range in maximum dimension from fractions of an inch to several inches. The quartz monzonite is highly jointed and weathers to angular rubble (fig. 3.45).

Looking west across the open area, one can observe the distinctive angular topography that has developed on this rock. It is undoubtedly related to the existence of vertically oriented, strongly resistant zones of silicification and to steeply inclined dikes which occur within the quartz monzonite. To the north, about 50 feet up the face, note a prospect tunnel which was driven through the original talus slope (now removed) in search of mineralization. The stock was mineralized locally along silicified fractures which were coated with pyrite, gold telluride minerals, fluorite, and

native silver. Careful examination of some of the larger blocks shows the nature of the mineralization in the form of silicified fracture fillings. Oxidation of the pyrite is visible in some of the blocks as brown stains.

To the south one can follow a small dirt road which rises to the entrance of the old Cutting Mine, operative during the early 1900's and developed for extraction of silver and gold at that time. About half way up the road there is a thin dike, 3-4 feet thick, of dark trachyte porphyry, rich in lath-shaped feldspar phenocrysts which are generally parallel to the contacts of the dike. Particularly dark, tabular zones within the dike represent locations of intense alteration along fractures. The dike is composed of well-developed orthoclase and

Figure 3.46. Photomicrograph of quartz monzonite from the Cutting Stock. Note altered orthoclase phenocryst (upper left), zoned plagioclase (P), and quartz (Q). Crossed nicols, 25X.

aegirine (sodic pyroxene) phenocrysts set in a fine-grained groundmass.

Stop 11.   Terry Peak

Location. Terry Peak, at an elevation of 7,064 feet above sea level, is the highest point in the northern Black Hills. It approaches the elevation of Harney Peak, the highest peak in the entire Black Hills region. Access to Terry Peak is obtained by taking the Terry Peak Chairlift from Nevada Gulch Road which intersects U.S. Highway 85 and U.S. Highway 14A approximately one mile south of Lead (fig. 3.47). Alternatively, one can drive directly to the top of the peak. From the town of Cheyenne Crossing, travel 2.6 miles east on U.S. Highway 85 and U.S. 14A to a junction with an unpaved forest service road leading to Terry Peak. Turn left (north) on the forest service road and travel to the top of the Peak.

History. Terry Peak was named for General Terry who was a prominent Union general and famous Indian fighter. In 1876 he was in command of the troops who were sent west to suppress the Sioux Indians and to return them to their reservations. His expedition moved into the Powder River-Yellowstone region of Montana to the northwest. The ill-fated Seventh Cavalry was

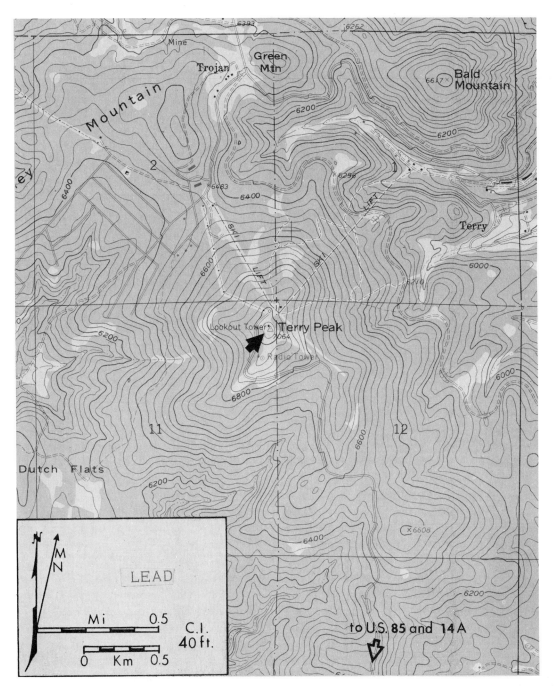

Figure 3.47. Stop 11, Terry Peak.

detached from his command, and, under the command of General Custer, this unit was separated from the main body of troops in Custer's fatal march to the Little Big Horn on June 25, 1876.

Description. From the top of Terry Peak, the panoramic view provides an opportunity to observe many of the major structural features of the Black Hills Uplift. Directly north is the Great Plains region, underlain by flat-lying sedimentary rocks which entirely surround the Black Hills. In the midground and separated from the main portion of the Black Hills by a broad valley, is a distinctive ridge, known as the Dakota Hogback, which forms the outer margin of the Black Hills Uplift (fig. 2.3). Separating the Hogback from the main part of the Black Hills is the Red Valley, also known as the Racetrack, which surrounds the entire Black Hills region, although best developed on the north and east margins. Inside the Racetrack, again looking north, one can see Elkhorn Peak, a nearly circular, dome-shaped area near the town of Whitewood. This region was domed by intrusion of igneous material which arched the overlying sedimentary rocks; erosion has not yet exposed the igneous rocks which underlie the area. Also visible to the north is Orman Dam, one of the first reclamation projects initiated by the United States to conserve water resources. Its construction was completed by the administration of Theodore Roosevelt.

By turning clockwise and following the skyline eastward and then southward, one can observe that the Black Hills Uplift is generally an elongate dome oriented with its long axis (about 120 miles) north-south and its short axis (about 60 miles) east-west. Directly east of Terry Peak, one ob-

serves the Great Plains again. In the distance, rising from the Great Plains beyond the rim of the Dakota Hogback, is a prominent hill known as Bear Butte (fig. 2.3). Like Elkhorn Peak, this feature is the result of doming of younger sedimentary rocks by the intrusion of igneous material which, in this case, is now well exposed in the upper portion of the Butte. Also to the east, but directly below Terry Peak, one can observe the town of Lead and the Homestake Mine in Lead. The general area of Lead and Terry Peak constitutes part of another upwarp referred to as the Lead Dome. Like Elkhorn Peak and Bear Butte, this area was domed by the force of igneous intrusive activity which pushed up the overlying sedimentary rocks. These igneous rocks are well exposed on Terry Peak and in the open cut in Lead. Localization and concentration of some of the gold mined at the Homestake Mine is thought to have occurred in association with this igneous activity.

Looking directly south one can observe Harney Peak in the distance (fig. 2.3). The highest point in the entire Black Hills Uplift, Harney Peak has an elevation of 7,242 feet above sea level. Harney Peak is underlain by resistant granite, characteristic of much of the ancient Precambrian terrain in the southern Black Hills.

The topography west of Terry Peak contrasts with that to the east. The upper surface of the Black Hills on the west is about 800 feet below the top of Terry Peak. It appears to be a rather broad, flat area (Limestone Plateau, fig. 2.3) dissected by a few streams following very narrow, deep canyons such as Spearfish Canyon. The upper surface, almost completely preserved on the western margin of the Black Hills, is underlain by the Pahasapa Limestone of Mississippian age. By contrast, this limestone has been removed by extensive stream erosion on the east side of the Black Hills, exposing the ancient Precambrian rocks below. Few broad upland surfaces remain on the east side which is dominated by steep slopes reflecting the steep dip of beds on the east side of the Black Hills in contrast to their relatively gentle dip on the west side. As a result, streams eroding the east side of the Black Hills have cut downward more rapidly and broadened their valleys, resulting in greater removal of the younger sedimentary rocks on this side of the Black Hills.

As figure 3.48 shows, the rocks immediately underlying Terry Peak are relatively young Cenozoic igneous rocks which have intruded Cambrian and younger Paleozoic sedimentary rocks. Other portions of this, or separate but related intrusions, are exposed sporadically in the vicinity of Terry Peak. The intrusion which underlies the Peak is thought to be a laccolith (fig. 3.48). As is typical of these Cenozoic intrusions in the Black Hills, the Terry Peak igneous rock is highly fractured into blocks which are strewn across the upper part of the Peak. Examination of a number of these blocks shows that the rock is composed of some larger mineral grains, approximately five millimeters in maximum dimension, enclosed in a groundmass of somewhat smaller grains, roughly one millimeter in maximum dimension. Microscopic study shows that this rock is a quartz monzonite porphyry composed of phenocrysts of zoned orthoclase, sodic plagioclase, and aegirine (sodic pyroxene) set in a fine-grained groundmass of quartz and feldspar (fig. 3.49).

Stop 12. Precambrian Rock Units in the Lead Area

Location. South and west of downtown Lead, a series of roadcuts provides an excellent opportunity to examine the Precambrian rocks which occur at the northernmost tip of the Black Hills (fig. 2.5). Because of their close proximity, these roadcuts are discussed as a group, and the specific location of each is given below. Additionally, each is keyed to figure 3.50.

Description. Although the garnet zone is represented east of Lead, all the rocks described here occur within the biotite metamorphic zone (fig. 2.5) which records the lowest intensity of metamorphism in the Black Hills. Several characteristics of these rocks reflect such conditions, namely (1) they are primarily fine-grained schists or finer-grained phyllites, (2) they contain micas and chlorite which form under low pressure/temperature conditions, and (3) they are characterized by relict features (inherited from the original rocks) such as bedding and sedimentary textures.

Detailed studies of the Precambrian schists have been made in the Lead area because of the critical nature of the schists with respect to gold occurrences. As a result, a number of distinctive mappable units have been recognized in this region (fig. 3.51). Particularly prominent are the Ellison, Homestake, and Poorman formations. Gold is restricted to the Homestake Formation.

At locality 12A (fig. 3.50), just north of the Terry Peak Turnoff on U.S. Highway 85, one can observe rocks that belong to the Ellison Formation. The rocks include

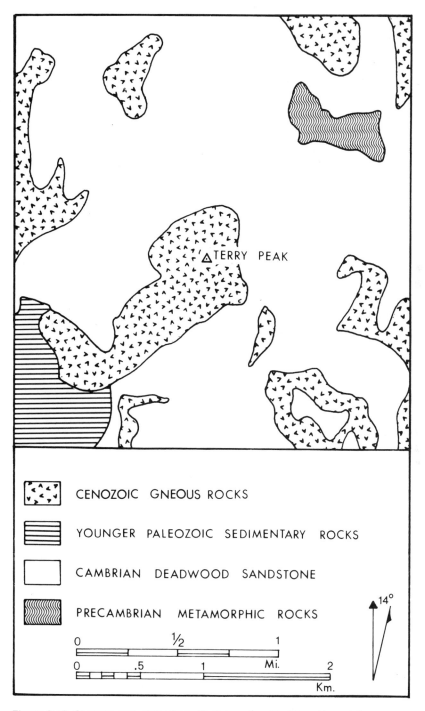

Figure 3.48. Geologic map of the Terry Peak area (modified from Homestake Mining Company unpublished map).

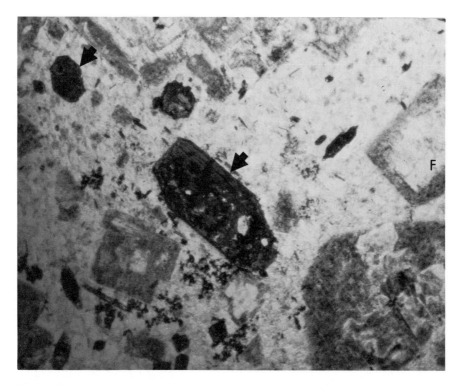

Figure 3.49. Photomicrograph of quartz monzonite porphyry from Terry Peak. Note the altered and zoned feldspar phenocrysts (F) and dark aegirine phenocrysts (arrow) set in a finer-grained quartz-feldspar groundmass. Crossed nicols, 25X.

dark gray quartzite, occurring as essentially vertical beds several inches to several feet thick, interlayered with gray to green platy phyllite (fig. 3.52). The quartzite is notable because of its relict sandy texture (fig. 3.53). At other localities one can observe relict cross-bedding and graded bedding in similar rocks. These features aid in interpreting the nature of the original sandstone from which the quartzite was derived. In addition, thick veins of quartz occur here as tabular bodies and as lenses which show prominent pinch-and-swell structure.

Further northeast along the exposure one can observe several thin intrusions of Cenozoic age. The thickest intrusion is a sill-like body (fig. 2.13), about 6-8 feet thick,

which weathers to a chalky white color. The rock is termed felsite in hand specimen; however, under the microscope it exhibits the composition of rhyolite. In the center of the body is a block of phyllite around which this body appears to have intruded. The body is cut by a large number of closely spaced fractures and joints which result in its break up into angular fragments, typical of the Cenozoic rocks in this region. The intrusion is probably the finest grained of these types of rocks in the area; it breaks with a conchoidal fracture.

Locality 12B (fig. 3.50) is further north on U.S. Highway 85 along the east side of the road. The cut exposes three rock units. The northernmost and southernmost units

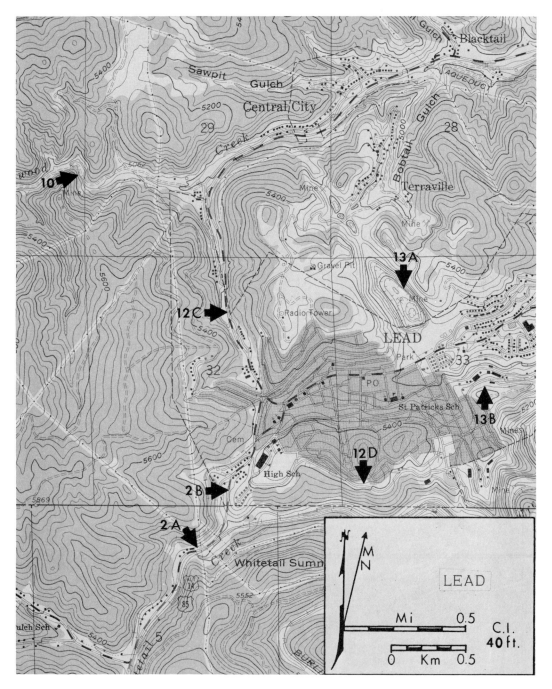

Figure 3.50. Stop 12, Lead area, South Dakota.

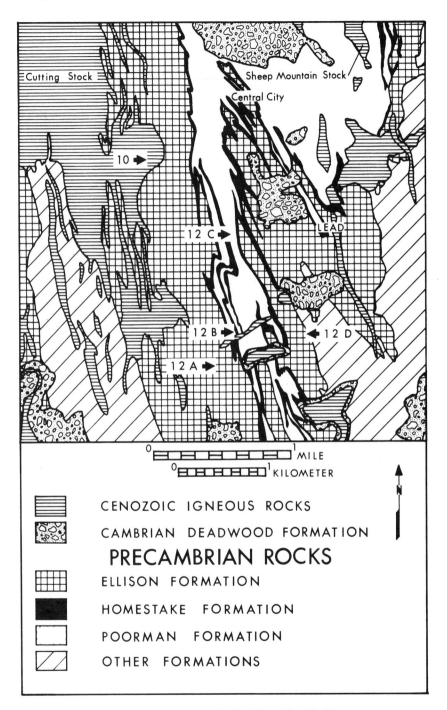

Figure 3.51. Generalized geologic map of the Lead area (modified from unpublished map, Homestake Mining Company.

Figure 3.52. Interlayered massive quartzite beds and phyllite, Ellison Formation.

Figure 3.53. Photomicrograph of quartzite from the Ellison Formation, showing relict clastic quartz grains in a fine-grained matrix. Crossed nicols, 25X.

Figure 3.54. Well-layered Poorman Formation (central two-thirds of photo) flanked by schist typical of the Homestake Formation.

are fine-grained schist typical of the Homestake Formation, whereas the middle unit is well-layered Poorman Formation (fig. 3.54). As is common, schists of the Homestake Formation here are red to brown (because of weathering of their iron-bearing minerals) and locally green due to the presence of chlorite. The main minerals of the Homestake Formation are quartz and sideroplesite (an iron-magnesium carbonate mineral), as well as biotite, chlorite, and graphite (Slaughter 1968). At this locality the Homestake Formation contains large quartz pods; recrystallized quartz pods and veins are common in the unit. As at locality 12A, the relict bedding is vertical here.

Locality 12C (fig. 3.50), directly opposite the entrance to the Ponderosa Motel on U.S. Highway 14A west of Lead, provides a good view of the Poorman Formation, the oldest Precambrian rock in the Lead area. The Poorman Formation is distinctive because of its very thin, well-developed, relict bedding (fig. 3.55). The rock is a gray phyllite composed of quartz, muscovite and/or biotite, graphite, and typically ankerite, an iron carbonate mineral (Slaughter 1968). At this locality a major fold can be defined if one observes carefully the inclination of the relict beds. They are almost horizontal at one point and then essentially vertical at the nose of the fold (fig. 3.56). Two episodes of folding have

Figure 3.55. Closeup of a representative slab of phyllite from the Poorman Formation, showing its characteristic relict bedding (color banding) and a well-defined lineation (crinkling).

Figure 3.56. Large fold in the Poorman Formation defined by the phyllitic (platy) structure which, near the figure, is inclined steeply away from the viewer but, near the top of the photo, is almost horizontal.

deformed this rock. The relict bedding is clearly visible and defines one period of folding. A well-defined crinkling which cuts cross the rock (fig. 3.55) was formed in a second episode.

The erosional boundary between these ancient Precambrian rocks and the immediately overlying, considerably younger, Cambrian rocks can be observed along Houston Street on the western perimeter of Lead (locality 12D, fig. 3.50). Travel west on U.S. Highway 85 and turn left (south) toward the high school. Travel uphill and continue to the road cut which is 1500 feet beyond the second road intersection after the high school.

The erosional boundary (unconformity) here is a gently undulating, sub-horizontal surface (fig. 3.57) which separates schist of the Precambrian Ellison Formation below from well-bedded sedimentary rocks of the Cambrian Deadwood Formation above. At this locality, roughly 20 feet of the sedimentary rocks are exposed. The base of the Deadwood is shale in some places and conglomerate in others. Above these units are roughly 15 feet of massive quartz sandstone into which several gold prospect pits have been developed. Sandy dolostone forms the uppermost unit. Below these sedimentary rocks, the Ellison Formation is a bluish-gray, fine-grained schist with steeply inclined foliation (fig. 3.57) which is finely crinkled. The rock is cut here by a Cenozoic porphyritic phonolite dike characterized under the microscope by the presence of orthoclase, sodic plagioclase, nepheline, and aegirine (sodic pyroxene) phenocrysts and the absence of quartz.

At the western end of this series of road cuts, all the rocks and the unconformity itself are offset along a steeply eastward-dipping fault (fig. 3.58). A subsidiary, westward-dipping fault is also visible cutting the Deadwood Formation and meeting the main fault just above road level. Careful tracing of beds on both sides of the subsidiary fault reveals a total displacement of roughly 10 feet.

Stop 13.   The Homestake Gold Deposit

Location.   Although no mineralization is visible at the surface, aspects of the occurrence, mining, and processing of the Homestake gold deposit may be observed at two locations within the city of Lead (fig. 3.50). North of the main street in Lead is the open cut (locality A) which exposes some of the rocks with which the gold ore is associated and which represents the earlier method used in mining the ore. The presently active underground mine workings and ore-processing operations are south of the main street at locality B. Tours are conducted regularly through the surface workings here. Both localities are readily accessible and marked by numerous signs in Lead.

Description.   The Homestake gold deposit is confined to a distinctive metamorphosed sedimentary rock unit, referred to as the Homestake Formation, in an area where this rock is cut by numerous Cenozoic intrusions (fig. 3.51). The Homestake Formation was described at Stop 12 and may be seen at locality 12B. The Homestake Formation, and associated formations, are intensely deformed as evidenced by their occurrence as large folds (anticlinal or up-folds and synclinal or down-folds) which themselves have been refolded, resulting

Figure 3.57. Unconformity between the Precambrian Ellison Formation (platy rock) below and the Cambrian Deadwood Formation (horizontal beds) above. Houston Road in Lead.

Figure 3.58. Steep easterly inclined fault (center of photo) which offsets the Precambrian-Cambrian unconformity. Note westerly inclined subsidiary fault (extending upward to right from figure). Houston Road in Lead.

in a contorted map pattern for these rocks (fig. 3.51). In addition to this later cross-folding, some bending of the rocks is related to forceful intrusion of the Cenozoic igneous rocks, particularly the dome associated with the Cutting Stock which is exposed west of Lead (Slaughter 1968).

Some of these relationships are well displayed in the open cut at locality A (fig. 3.59). This pit was the site of the original surface mining and subsequent underground mining of the gold deposit. The rock which was mined, the Homestake Formation, is now lacking in the pit which has collapsed considerably since the early mining. Presently exposed in the pit is the Poorman Formation on the east side and the Ellison Formation on the west side. Particularly on the east side of the pit (fig.

3.60) one can observe a series of rhyolite sills and dikes and part of a quartz monzonite sill (at the north end of the cut) as shown in figure 3.60. The quartz monzonite sill exhibits well-developed columnar jointing, particularly at its base where it is in contact with the well-bedded Deadwood Sandstone. The progressive elevation of the Deadwood Sandstone constitutes clear evidence of the forceful injection of these intrusive bodies which have generally domed up this area.

Originally a mountain, the open cut marks the site at which the Homestake claim was located on April 9, 1876. The present pit is 500 feet deep, 1300 feet wide, and 4300 feet long. Since 1945, when use of the open cut ceased, all mining has been conducted by means of underground meth-

Figure 3.59. Open cut mine in Lead north of the present mine and ore processing facilities.

Figure 3.60. Panorama of the east side of the open cut in Lead. Right two-thirds of view shows rhyolite sills (light-colored bands) and some dikes within the Poorman Formation (dark zones). Left one-third of view shows a quartz monzonite sill (underlying the peak) with vertical columnar joints, underlain by horizontal beds of altered Deadwood Sandstone, in turn underlain by the (darker) Poorman Formation.

ods. The site of this activity may be viewed by looking southeast of the open cut toward the hill south of Lead (fig. 3.61). The ore processing facilities are located on top and along the side of the hill below a prominent structure, the headframe of the Yates shaft, one of two major shafts which provide access to the underground workings and to the ore.

The ore deposit consists of a series of elongate, pencil-like bodies which occur in the Homestake Formation. Most of the ore is localized near the boundary between the biotite and the garnet zones of metamorphism (fig. 2.5); where the Homestake Formation changes from cherty quartzite and sideroplesite (magnesium-iron carbonate) schist to quartzite and cummingtonite (iron-magnesium amphibole) schist (Norton and Redden 1975). Mineralized portions of the Homestake Formation, of either type, are typically altered to chlorite and are rich in quartz veins (fig. 3.62). The common metallic minerals are native gold (20% silver alloy), arsenopyrite (arsenic-iron sulfide), pyrrhotite (iron-nickel sulfide), and pyrite (iron sulfide). Although most of the gold is microscopic in size,

some is clearly visible with the unaided eye. The average gold content of the ore is roughly one-third of an ounce, whereas the silver content is 1/25 of an ounce per ton (Slaughter 1968). A synthesis of studies by Slaughter (1968), Rye and Rye (1974), and Rye and others (1974) indicates that the main interval of gold deposition occurred during Precambrian time.

Mining. The ore is extracted by means of an intricate network of some 200 miles of underground tunnels serviced by two shafts—the Yates shaft and the Ross shaft (fig. 3.63). There are 40 active mine levels (53 total) spaced 150 feet apart. At present the deepest point in the mine is at 8,200 feet below the surface, and development is presently occurring as deep as the 8,000 foot level. The surface elevation of the two shafts is approximately 5,200 feet above sea level. Thus, mining is presently in progress well below sea level in this mountainous area. In the Yates shaft there are two skips and two cages which travel vertically at up to 35 miles per hour. The pair of skips is for ore extraction, and the cages are for movement of men and equipment. Their

Figure 3.61. View southeast from the Lead open cut, showing the Homestake ore milling facilities and the headframe of the Yates Shaft.

Figure 3.62. Closeup of typical Homestake gold ore showing white quartz, silvery metallic arsenopyrite, and dark chlorite-rich schist.

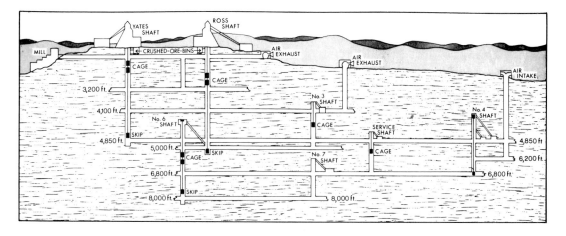

Figure 3.63. Generalized cross section showing workings of the Homestake Mine (modified from Homestake Mining Company 1976).

movement is controlled in the hoist room which contains huge tapered drums on which cables, slightly less than two inches in diameter, are wound as skips or cages are raised or lowered. The mine employs approximately 1300 people, of whom roughly 1,000 work underground, including 100 women.

The ore is extracted largely by means of the cut-and-fill method (fig. 3.64) and, to a lesser extent, by several other methods. As shown in Figure 3.64, the cut-and-fill method involves developing a stope or opening in the ore by continually blasting and collapsing the roof as the floor of the stope is elevated with sand backfill provided from the ore refining process. The broken ore is scraped, by means of a slusher or drag shovels, to a chimney which is built and extended as the stope is backfilled. The material from each stope is transported by rail cars to skip pockets for hoisting to the surface. Here it is crushed to fragments having a maximum diameter of four inches and then stored temporarily underground.

Milling. Processing of the ore begins wtih fine crushing that reduces it to fragments 1⅛ inch in diameter, followed by screening, and then crushing to 3/8 inch diameter. It then proceeds to a rod mill (fig. 3.65) which reduces it to the consistency of wheat by means of steel rods which revolve with the ore in a moving cylindrical drum. It is then fed to a ball mill where it is pulverized to the consistency of flour by means of steel balls which move in a drum with the ore. Ultimately, the ore powder is subjected to gravity concentration and cyanidation. The latter process involves dissolving the gold and silver in a weak solution of sodium cyanide. The metals are then recovered from this solution by addition of zinc dust, or activated carbon. Further refining by smelting and treatment with chlorine gas removes the remaining impurities and facilitates separation of the gold and silver. From the time the ore is removed from the mine to the time that the gold and silver are extracted from it and waste returned to the mine, approximately two weeks elapse.

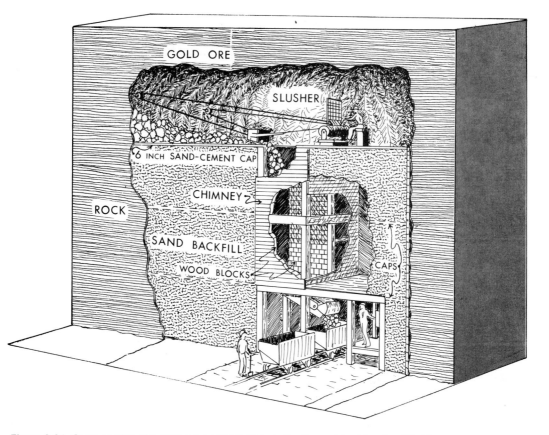

Figure 3.64. Cross section of a typical cut-and-fill stope in the Homestake Mine (modified from Homestake Mining Company 1976).

When the price of gold was $35 per ounce, the Homestake Mining Company prepared 35-lb. gold bars which were sold directly to the U.S. Government. At the present inflated price of gold, the Company produces 20-lb. gold bars which are transported to the Rapid City airport from whence they are shipped to San Francisco and ultimately sold to private companies. The Homestake Mining Company produces approximately 70 pounds of gold and 20 pounds of silver per day.

Stop 14. Strawberry Ridge Iron Deposit

Location. This deposit is several hundred feet west of U.S. Highway 385, roughly 1500 feet northwest of the Strawberry Hill Campground (fig. 3.66), and approximately five miles southeast of Deadwood. Turn west off the highway onto a dirt road, park, and walk into the open pit.

Description. This mine exposes iron concentrated near the base of the Cambrian

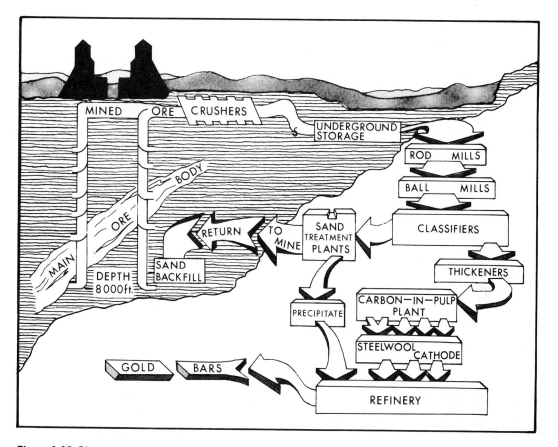

Figure 3.65. Diagrammatic representation of steps in the processing of ore extracted at the Homestake Mine (modified from Homestake Mining Company 1976).

Deadwood Formation, a common occurrence in the Black Hills region. Careful examination of the pit shows that its floor is developed essentially in bluish to reddish Precambrian iron-bearing schist and related rocks.

Immediately overlying the Precambrian rock is a relatively massive unit which constitutes the base of the Deadwood Formation here. This unit, 6-8 feet thick, consists of very coarse red conglomerate, sandstone, and shale, all rich in iron oxides (fig. 3.67).

The main iron oxide mineral is red hematite which occurs as cementing material in the conglomerate and sandstone, as fine-grained disseminated material in the shale, and as pebbles and cobbles in the conglomerate (fig. 3.68). Some of the pebbles consist of iron-rich rock not greatly dissimilar to the Precambrian rock exposed in the floor of the pit. According to Harrer (1966), holes drilled up to 200 feet beyond the limit of the pit indicate that the hematite-rich basal unit of the Deadwood For-

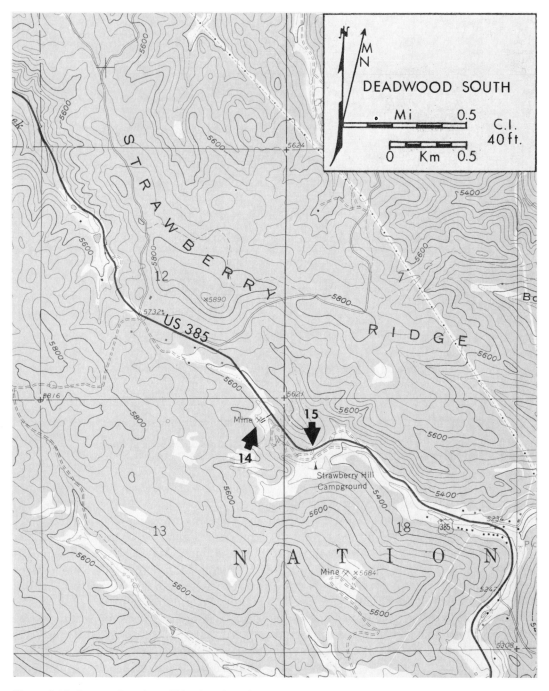

Figure 3.66. Stop 14, Strawberry Ridge iron deposit.

Figure 3.67. Strawberry Ridge open pit exposing well-bedded sandstone (background) underlain by iron-rich sandstone and conglomerate (near 4 figures), underlain by Precambrian schist.

Figure 3.68. Closeup of iron-rich basal conglomerate of the Deadwood Formation and thin Cenozoic dike (white) to rear (arrows).

mation is continuous in this area. At one location (fig. 3.68) a one-foot thick, steeply dipping dike of light-colored, very fine-grained Cenozoic igneous rock cuts across the conglomerate.

The uppermost unit of the Deadwood Formation exposed at this site consists of approximately 15 feet of typically buff-colored, well-bedded sandstone which immediately overlies the basal unit (fig. 3.67). By contrast with the basal unit, this unit is lacking in iron except for one or two local pockets, a foot or two across, which are distinctively red and apparently rich in hematite. These pockets are clearly above the top of the basal iron-rich unit previously described.

Regarding origin, the basal iron-rich unit appears to be the result of weathering of the iron-rich Precambrian rocks. At this particular locality, the well-rounded character of the iron-rich pebbles suggests that much of the material here was transported from another source area where Precambrian iron-bearing rocks, similar to those exposed on the pit floor were undergoing weathering and erosion. The pockets of hematite which are scattered through otherwise iron-poor sandstone in the upper unit here may be related to hot fluids associated with Cenozoic intrusive activity which mobilized iron from below and transported it upward along favorable, relatively permeable zones within the sandstone. An alternative, although less likely, is that the source of this local iron was from a higher stratigraphic unit within which it was mobilized and transported downward into this standstone by groundwater.

According to Harrer (1966), several thousand tons of hematite-rich material have been extracted from this and other nearby pits. Its main use has been as a source of mineral pigment and as an additive for the manufacture of cement at the State Cement Plant in Rapid City. The typical material in the basal unit averages 43% iron. However, selected samples contain up to 66% iron (Harrer 1966).

Stop 15.   Strawberry Ridge Schist

Location.   The schists in the Strawberry Ridge area are well exposed in this roadcut along U.S. Highway 385, approximately five miles southeast of Deadwood (fig. 3.66). The cut is on the north side, along a sharp curve in the road, directly opposite a narrow dirt road to the south.

Description.   The cut exposes metamorphic rocks which were formed under conditions slightly more extreme than those which prevailed in the area described at Stop 12, west of Lead, to the north of this locality. Note the location of this area by comparing figures 1.3 and 2.5. The rocks here occur within the northernmost of two garnet zones which flank the biotite zone in the Lead area. As expected, the rock is slightly coarser grained than most of the metamorphic rock west of Lead. It is a fine- to medium-grained quartz-mica schist which contains small pink garnets. Although subjected to more intense recrystallization than the rocks west of Lead, this rock also exhibits relict bedding, expressed by different colored layers from one-half to several inches thick. The bedding testifies to its sedimentary parentage (fig. 3.69).

The layers have been folded into a series of open folds which plunge steeply toward the road (fig. 3.69) and which themselves are cut by well-defined schistosity, locally parallel or sub-parallel to the relict bed-

Figure 3.69. Relict bedding (color banding) in schist deformed into steeply plunging folds.

ding but otherwise cutting it at a steep angle. In addition, the rock has been deformed into a series of minor folds which show up particularly well in muscovite-rich layers that are crinkled. Finally, the rock exhibits excellent boudinage, or "sausage-link" structure, defined by what was originally a continuous layer of quartzite that has now been pinched into separate segments. Careful examination allows one to trace this feature across a span of six feet in the center of the roadcut. Note that this is part of a single large fold. The schist here differs from the Precambrian schist exposed in the quarry floor at the Strawberry Ridge iron deposit (Stop 14) several hundred feet to the northwest (fig. 3.66). This schist contains iron; however,

its iron content is less than that in the schist exposed in the quarry.

Although the main part of the roadcut exhibits steeply plunging, smoothly curving, "cylindrical" folds, the uppermost portion, particularly at the west end, shows that the schist has been deformed into very tight, "angular," chevron folds (fig. 3.70). Such folding is the result of modern surface gravity effects referred to as "creep."

Stop 16. Cenozoic Volcanic Rocks

Location. This unique locality is on the property of the Tomahawk Lake Country Club situated east of U.S. Highway 385 roughly seven miles southeast of Deadwood (fig. 3.71). East of the clubhouse, follow

Figure 3.70. Panorama of roadcut showing relict bedding in schist, steeply plunging folds, and (in uppermost part of the cut) chevron folds.

the dirt road along the east side of the golf course and turn off to the right near a large pine tree where a small valley intersects the main open valley. Progress eastward up the small valley several hundred feet to an exploratory cut (locality 16A, fig. 3.71). Then climb almost due north up to the crest of the small hill (locality 16B, fig. 3.71) to the old mine timbers. At the crest of the hill is an open, abandoned mine shaft (be careful to avoid proximity to the shaft) and another tunnel-like cut on the side of the hillcrest.

Description. At the base of the hill, the exploratory cut (locality 16A) exposes a basal unit of thinly bedded, steeply dipping, lithic tuff (Kircher 1977), a volcanic rock consisting of very small rock fragments, less than ⅛ inch in diameter, associated with explosive eruption. Immediately overlying this fine-grained rock is a much coarser volcanic breccia containing angular fragments up to three inches across. Interestingly, fragments of fine-grained volcanic rock are interspersed with (Precambrian) schist fragments in these rocks.

At the crest of the hill (locality 16B), the rock consists of one or two masses of volcanic glass (ranging from black obsid-

ian to brownish pitchstone), which are interpreted as large inclusions (Kirchner, 1977) within the second rock type. This rock type is a generally pinkish volcanic breccia made up of an assortment of very fine-grained rock fragments up to 2 inches in diameter (fig. 3.72). According to Kirchner (1977) flow layers in one of the masses of volcanic glass are truncated on all sides by their contact with the volcanic breccia. With great care, one can look down the abandoned mine shaft and observe a vertical contact between the glass and the breccia as well as a horizontal contact (about 15 feet downward in the shaft). The small cut on the side of the hillcrest affords an easier view of another contact between the two rocks (fig. 3.73).

Close examination of the glass shows that it is characterized by an intricate, anastomosing network of curving fractures, some of which appear to be silicified. In addition, the rock contains scattered crystals (phenocrysts) which indicate that the molten material from which it formed began crystallizing slowly at depth before it was suddenly erupted onto the surface where it chilled rapidly to a glass. The associated volcanic breccia in this deposit represents explosive eruption of molten rock

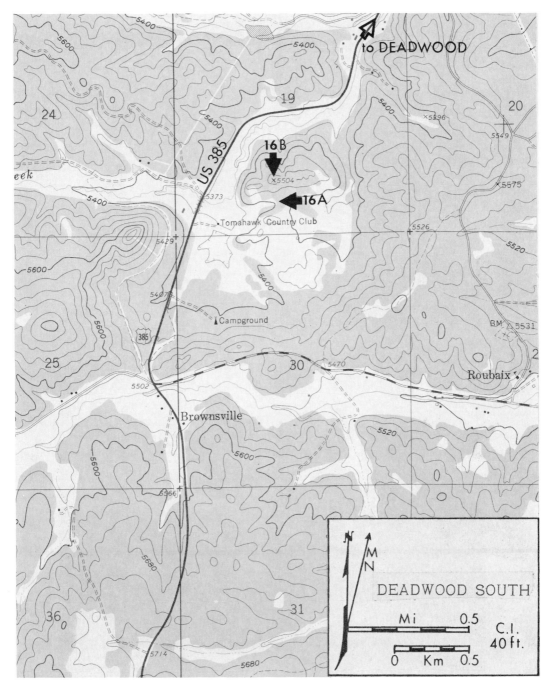

Figure 3.71. Stop 16, Cenozoic volcanic rocks.

Figure 3.72. Closeup of volcanic breccia showing heterogeneous size of angular rock fragments.

Figure 3.73. Steep contact (arrows) between volcanic breccia (left) and glass (right).

material which solidified into angular, fine-grained (aphanitic), fragments before they settled to the ground.

As indicated in Chapter 2, these extrusive, or volcanic, rocks are significant because this is the only known locality in the Black Hills where Cenozoic-age magmas reached the surface. In addition the age of this rock (10.5 million years) extends the range of Cenozoic igneous activity to the Miocene Epoch (Kirchner 1977), a considerably more recent interval of time than the Eocene age (38.8-60.5 million years back) for the Cenozoic intrusive activity.

Stop 17. Nemo Iron Deposit

Location. This locality consists of a small rock exposure along the west side of the Nemo road roughly 0.4 miles south of Nemo (fig. 3.74). The rock crops out at road level as a number of jumbled blocks at the point where a row of pine trees begins on the east side of the road.

Description. The rock exposed here, referred to as taconite, is an excellent example of typical iron formation in that it consists of well-defined, thin layers of iron oxide minerals alternating with siliceous layers (fig. 3.75). Iron formation occurs on all of the continents, and, in the United States, major commercial deposits exist in the Lake Superior region. Such rocks are interpreted to have been deposited as chemical sediments originally. Because most iron formation is geologically quite old, most of it has been recrystallized due to metamorphic activity since its initial deposition.

In the Nemo iron formation, the iron minerals are confined largely to fine-grained bluish layers up to ½-inch thick and separated from one another by white to yellowish quartz-rich layers typically ⅛-inch thick. In general, the layering is oriented vertically here, but locally in this exposure it is intensely deformed into a series of very tight, small folds. The iron formation is relatively resistant in this area and defines a ridge which extends northwest from this point. The rock can be traced to the northwest and also farther to the southeast where it rises as a ridge 200 feet above Boxelder Creek. At that location it is 100-300 feet thick (Harrer 1966).

In this region the iron formation varies from a fine-grained, layered, iron oxide-rich quartzite to a very siliceous and iron oxide-rich slate and schist (Harrer 1966). The iron oxides consist of specular hematite, martite (hematite which has replaced magnetite), and magnetite (fig. 3.76). According to Harrer (1966), the Nemo iron formation averages typically 30-42% iron.

In general this is the same kind of material which forms the floor of the pit at the Strawberry Ridge hematite locality. However, at that locality the iron-rich Precambrian rock is largely schist, whereas here it is largely quartzite. At both localities the iron formation is immediately overlain by the Cambrian Deadwood Formation. In the Nemo area, these Cambrian rocks are clearly visible to the east across Boxelder Creek where they exhibit horizontal bedding and contain much red material. Across the Creek one can see a quarry developed in the iron formation. The material is extracted as an iron additive for mixing with

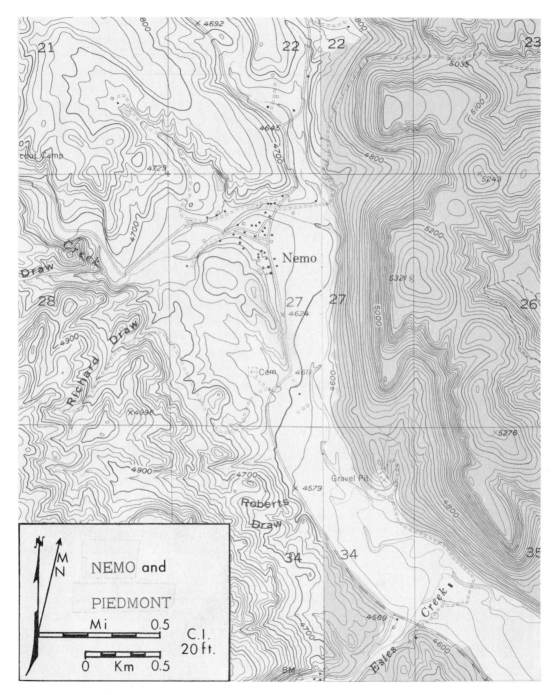

Figure 3.74. Stop 17, Nemo iron deposit.

Figure 3.75. Closeup of well-layered iron formation along the road just south of Nemo.

Figure 3.76. Photomicrograph of Nemo iron formation showing granoblastic quartz layers alternating with layers of opaque iron oxides. Crossed nicols, 25X.

cement manufactured at the State Cement Plant in Rapid City.

Stop 18. Little Elk Creek Canyon—White Gate

Location. About two miles west northwest of Piedmont, South Dakota. Access to this locality is via a gravel road just north of exit 44 on U.S. Highway I-90. Exit west, turn north at the dead end immediately adjacent to the exit, and travel 0.2 mile to a gravel road leading toward the hills. Follow the road until it ends, about 1.5 miles, and walk along the old road bed to White Gate (fig. 3.77).

Description. Two completely different sets of observations can be made in the area of White Gate. The first relates to the geologic processes that have deformed the bedrock in this area; the second relates to the violent stream activity which has occurred in the valley of Little Elk Creek.

The rocks encountered in traversing the valley toward and through White Gate include, in order, the Minnelusa Formation, the Pahasapa Formation, the Englewood Formation, and the Deadwood Formation. As one travels around the margin of the Black Hills and observes the escarpments leading toward the inner Hills, it is clear that the beds dip more or less uniformly 20-30° away from the Black Hills. In fact, as one enters Little Elk Creek Canyon it can be observed that the Minnekahta, Opeche, and Minnelusa rocks dip east at about 20° (fig. 3.78). However, as one approaches White Gate, the dip steepens markedly, so much so that at White Gate the dips approach vertical (fig. 3.79). This

remarkable increase in dip of the Paleozoic rocks defines a relatively localized geologic structure, the Little Elk Creek Monocline. A monocline is a folded structure in which the beds dip uniformly in a single direction over a relatively short distance, then steepen in dip, and then, again, become more gently dipping. This monocline is of relatively local significance in that it can be traced north and south for a distance of only about five miles. A similar small monocline exists on the eastern flanks of the Black Hills in the area of Buffalo Gap, near Hot Springs. Much more significant monoclines, in terms of their extent, rim most of the western edge of the Hills but produce less dramatic structures than at White Gate. As one passes through White Gate and continues up the valley, the geologic section is traversed rapidly; within a distance of about 700 feet, one encounters the upper part of the Cambrian Deadwood Formation. This forms a structure similar to that seen at White Gate and is referred to locally as Red Gate. Just beyond Red Gate, as one travels a bit farther up the stream valley, the dips again become more gentle, but the beds continue to dip east. At this point one has traversed the Little Creek Monocline.

This exposure of Paleozoic rocks affords an excellent opportunity to examine the fossil content of the Pahasapa and the Deadwood formations. At the contact between the Pahasapa and the overlying Minnelusa Formation, best observed at creek level just downstream from White Gate, large colonies of the coral *Syringopora* are preserved in the Pahasapa Formation (fig. 3.80). Occasionally brachiopods can be found preserved in the rock between coralline masses. Farther up the

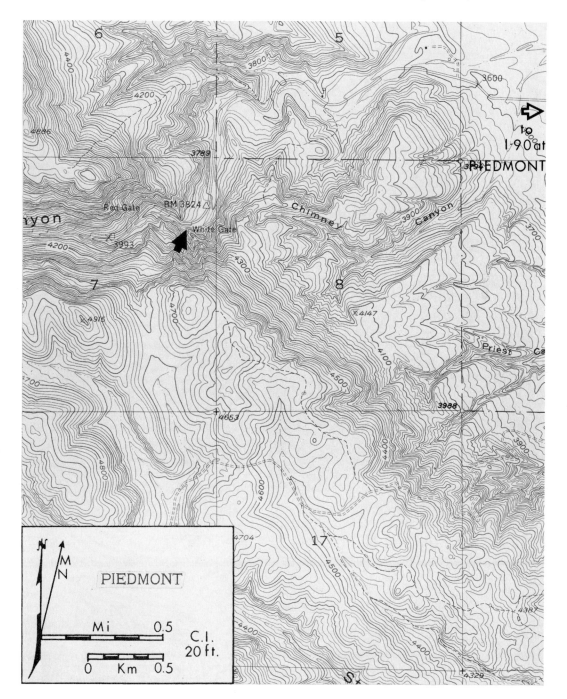

Figure 3.77. Stop 18, Little Elk Creek and White Gate.

Figure 3.78. Little Elk Creek Canyon viewed from the Red Valley. The prominent rock unit forming the dipping surface at the mouth of the canyon is the Minnekahta Formation. The boulders in the foreground were washed out of the canyon by flood activity.

Figure 3.79. The vertical wall in the right foreground is steeply dipping Pahasapa Formation of White Gate. Note that the beds exposed high on the hillside in the background dip much more gently.

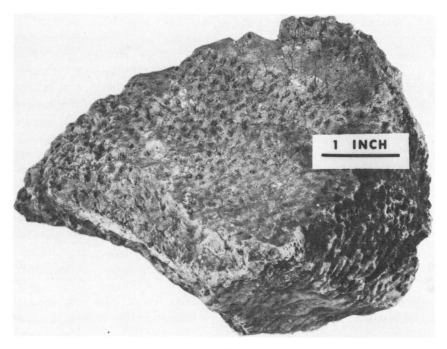

Figure 3.80. The coral *Syringopora* collected from near the top of the Pahasapa Formation.

stream on what used to be the road grade, the Pahasapa can be studied in detail, and numerous brachiopods and solitary corals can be seen and collected. In addition, some interesting sedimentary structures are visible in the formation at this point. For example, recalling that the bedding planes, the layers paralleling the surface on which the grains of calcium carbonate were deposited, are vertical, one can observe well-developed sets of cross beds. Individual units were deposited as a series of "dunes" that migrated across the sea floor in response to wave activity. The upper surfaces of these dunes were inclined to the horizontal, and their remnants can be seen as cross bedding.

Fossils can also be collected in the Deadwood Formation just beyond Red Gate. Here the formation tends to be silty shale;

by carefully observing the bedding planes on the shale and in the fine sandstone, tracks, trails, and burrows can be collected along with occasional fragments of triloobites. In fact, fossil collecting in the Deadwood Formation is as good here as it is anywhere in the Black Hills region.

Comparing the geologic section of the entire Black Hills Paleozoic rock units to the section observed along Little Elk Creek suggests that some of the units, notably the Winnipeg Formation and the Whitewood Formation, have been reduced in thickness or are missing altogether. In fact, the Englewood Formation can be demonstrated to exist in the area only by climbing well up on the sides of the hills over Red Gate. Exposures of the Englewood are nowhere to be seen at road level. This thinning or elimination of rock units in this particular area may be a result of reduction of thickness at the time of deposition, erosion of these rocks on the eastern flank of the Black Hills, or squeezing and "structural thinning" during the time of folding of the Little Elk Creek Monocline.

The second major set of observations that can be made in Little Elk Creek Canyon relate to dynamic stream processes. As recently as 1972 the road up Little Creek Canyon served as the main access road to Wonderland Cave located about three miles northwest of White Gate. In June 1972, heavy rains produced massive flooding on the east flank of the Black Hills. Although the drainage basin of Little Elk Creek (that is, the area in which stream water is collected by the Creek), is relatively small, enough water was collected to totally destroy the road in the area of

Red Gate and White Gate (fig. 3.81) and to move huge boulders weighing several tons downstream. The road has not been repaired subsequently. Examination of these boulders suggests that a terrific volume of water moved through the valley during this brief period of flooding. As one proceeds back out of the valley toward I-90, it can be observed that similar-sized boulders lie well out into the Red Valley (fig. 3.78). Most of these boulders were carried by earlier floods, but examination of the present Little Elk Creek valley shows that the stream is periodically capable of transporting large boulders. Under normal flow, little or no water is visible in the stream channel. These kinds of observations lead to the suggestion that much erosion in stream valleys of this sort occurs as spasmodic, nearly catastrophic, events

Figure 3.81. Boulders filling the valley of Little Elk Creek as a result of the flood in 1972. The road in this section of the canyon had previously been located along the right side of this valley.

and that the normal rate of erosion of the valleys, when water is flowing at its typical rate, does relatively little work.

Stop 19.   Dinosaur Overlook

Location. Dinosaur Park, Rapid City, South Dakota. Access to this locality from Main Street in Rapid City is via West Boulevard and Quincy Avenue. Travel south from Main Street three blocks on West Boulevard and turn right (west) on Quincy Avenue. Travel along Quincy 0.8 mile to Dinosaur Overlook, park, and walk to the top of the hill (fig. 3.82).

Description. Dinosaur Overlook provides a spectacular view of almost the entire east side of the Black Hills Uplift (fig. 3.83). The Overlook rests on the Lakota Sandstone which caps most of the "Dakota" Hogback surrounding the uplift. As one views the ridge to the north, the following rocks (below the Lakota Formation) can be identified in descending order: the Morrison Formation, the Unkpapa Sandstone, the Sundance Formation, and the redbeds of the Spearfish Formation which form the broad valley known as the Racetrack. The Racetrack can be traced from near the town of Sturgis, north through west Rapid City, and south for a few miles. These rocks of the Racetrack and the "Dakota" Hogback form the Mesozoic rim of the Black Hills Uplift.

Viewing the Hills themselves, the gentle slope bordering the main part of the uplift is developed on the Minnekahta Limestone of Permian age. Looking west-northwest one can see a large cement plant where the Minnekahta Limestone is being quarried from this ridge for use as a raw material in the manufacture of Portland Cement. Beyond the Minnekahta ridge, the main core of the Black Hills Uplift can be observed. The highest peak visible is Harney Peak, which lies southwest of the overlook and is the highest point in the Black Hills at an elevation of 7,242 feet. Harney Peak is composed of the Harney Peak Granite, an extensive Precambrian igneous rock unit best known because the Mount Rushmore Memorial was carved in it.

Looking slightly southwest one can see the canyon of Rapid Creek and can follow the trace of Rapid Creek through west Rapid City to the point where it passes through the Dakota Hogback. This stream is one of many drainages which radiate from the central portion of the uplift toward the margins (fig. 2.4). Rapid Creek and many of the other tributaries on the south and east sides of the Hills drain into the Cheyenne River which, in turn, drains into the Missouri River. North of this point most of the streams join the Belle Fourche River which drains into the Cheyenne River. Therefore, two types of drainage are typical of this region (fig. 2.4): a set of streams which drain from the core of the Hills toward the margins and are radial in pattern; and major streams, the Cheyenne and the Belle Fourche rivers, which are peripheral to the uplift and which are positioned because of the structure of the Black Hills.

Rapid Creek was the site of massive flooding in 1972. The dam at the mouth of Rapid Canyon, which serves as the chief water supply for Rapid City, was ruptured on June 9th (fig. 3.84) by excessive rainfall and an extremely rapid rise in creek level. The dam burst and water flooded a broad reach on either side of Rapid Creek resulting in millions of dollars in property

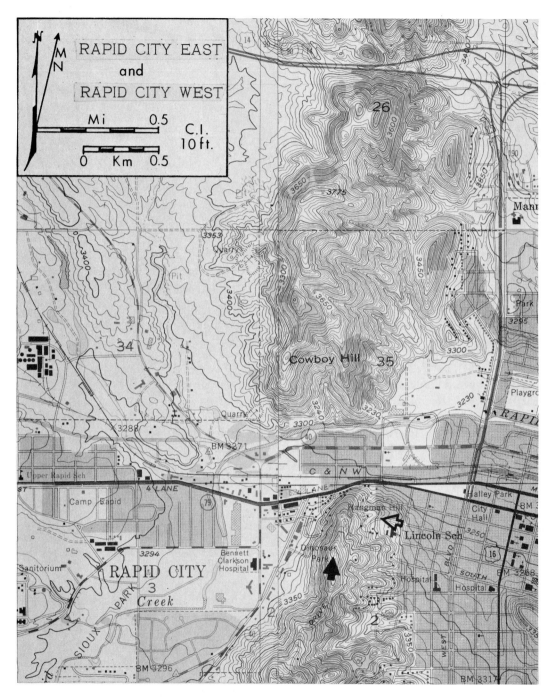

Figure 3.82. Stop 19, Dinosaur Overlook, Rapid City, South Dakota.

Figure 3.83. Panoramic views of the Black Hills from Dinosaur Overlook. The upper view is to the north and northwest, the lower view is to the south and southwest.

Figure 3.84. Rapid Creek Dam just after it burst in the flood of June 9, 1972.

Figure 3.85. Houses along Rapid Creek in western Rapid City that were destroyed by floodwaters of Rapid Creek.

damage (fig. 3.85) and the loss of many lives. Subsequent to this, the stream was channelized and a new dam was con-constructed as part of an effort to prevent further flood damage.

Viewing the area to the east of Dinosaur Overlook, one is able to see a broad reach of the Great Plains which, in this area, are predominantly underlain by the Pierre Shale of Late Cretaceous age. In the far distance almost directly east of the site, one can view, on a clear day, some of the flat-topped mesas which are underlain by the White River Formation in the Badlands National Monument.

Stop 20. Peerless (Rushmore) Pegmatite

Location. This pegmatite, known locally as the Rushmore Pegmatite, is well exposed at the Peerless Mine 0.5 mile south of Keystone. Travel west out of Keystone to U.S. Highway 16A. Turn left (south) and travel to Roy Street (at Mister Donut shop). Take Roy Street 0.4 mile, bearing right along the winding gravel road, to a "T" intersection. Turn left and proceed up the hill 0.1 mile to the mine (fig. 3.86). Permission to enter must be obtained from the owner, Mr. Paul Mitchell, who lives in the house on the hill top several hundred feet left (west) of the mine entrance. Use extreme caution within the mine.

History. According to Sheridan et al. (1957) the Peerless Pegmatite was mined initially in 1907, primarily for the lithium mineral amblygonite. Since then it has been developed for its beryl, scrap mica, sheet mica, potassium feldspar, plagioclase feldspar, tantalite-columbite, and cassiterite as well. In recent years beryl and scrap mica are the two major minerals that have been

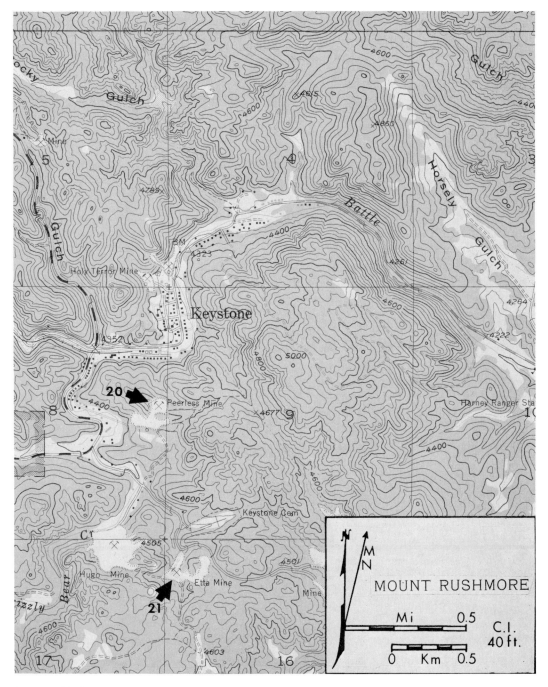

Figure 3.86. Stop 20, Peerless Pegmatite.

mined here. In 1977 the property was acquired by Mr. Mitchell who is developing it for visitation by the public.

Description. Although it is not easy to detect the zoning here without detailed and extensive examination, the Peerless Pegmatite is a good example of the commercial, zoned pegmatites in this region. According to Sheridan et al. (1957) it is a complex body composed of seven interior zones, two replacement units, and two types of fracture fillings (fig. 2.17). The various units are defined primarily on the basis of their differing proportions of quartz, plagioclase, microcline perthite, muscovite, and lithia mica. The mineralogic zones are generally comparable to those given in Table 2.2. The body is an irregular stock-like mass (fig. 2.13) which, in cross-section, has the shape of an elongate dome. In map view it has a tear-drop shape with its long axis oriented generally northwesterly. Its maximum map dimensions are 580 feet by 360 feet (Sheridan et al. 1957).

The body has intruded Precambrian schists, primarily quartz-mica schist composed of 45-65% quartz, 15-30% muscovite, 10-25% biotite, as well as small amounts of staurolite, garnet, and several other minerals (Sheridan et al. 1957). The presence of staurolite suggests that these rocks have undergone medium-intensity metamorphism. Note that the Keystone area lies within the staurolite zone on figure 2.5.

At the mine entrance one can observe directly ahead a number of distinctive pod-like bodies of dark quartz-mica schist which appear to be suspended down into the roof of the light-colored pegmatite. The more obvious minerals in this granitic pegmatite are microcline perthite (composed of microcline with thin films of plagioclase in it), quartz, and muscovite mica. The abundant irregular black areas are aggregates of tourmaline 3-4 inches in diameter scattered across the walls of the mine. To the right (and less obviously to the left) at the mine entrance, one can see a thin (2-3 feet thick) pegmatite dike with contacts that cut across the structure (platy character) of the adjacent schist. Composed essentially of the same minerals as the main body, this dike is probably an offshoot of the large pegmatite which was mined here.

Standing approximately at midpoint on the bridge over the main open cut (fig. 3.87), one can obtain a good overview of the features to be seen here. Northwesterly is the contact between the pegmatite and the schist (fig. 3.88) as well as a large amount of feldspar and muscovite which are well-exposed in the mine face. Looking northerly and northeasterly from the bridge one can see numerous sharp contacts between the pegmatite and irregular masses of schist as well as several extensions of schist tapering down into the roof. From the bridge one can also readily observe the several mining methods used here. The main method involved development of a glory hole below the bridge which was extended down the axis of the pegmatite. A glance at the mine walls reveals, in addition, a number of individual adits (or tunnels) along which a particular mineral was sought in the mining of pegmatite (fig. 3.87).

After leaving the bridge, rounding the curve into the first open area, and looking directly ahead toward the main open tunnel, one can see well-exposed contacts between the pegmatite and the schist on both sides of the tunnel entrance (fig. 3.89). Note that the pegmatite is generally much finer grained near the contact as compared

Figure 3.87. Aerial overview of the Peerless Mine showing the light-colored pegmatite and intervening areas of schist as well as the diverse group of tunnels, cuts, and pits used to mine the body. Note the bridge (arrow) referred to in the text.

Figure 3.88. Northwesterly view from the bridge in the Peerless Mine, showing a sharp contact between the pegmatite and schist.

Figure 3.89. Detail of a sharp contact between the Peerless Pegmatite (left) and schist.

with farther from it. Notice also that the pegmatite contact cuts across the steeply inclined schistosity of the darker rock. Overhead one can see a series of irregular fractures and some parallel joints which characterize the pegmatite. The steep inclination of the schistosity is obvious as one walks along the main tunnel to the large opening at the end of the tunnel. Notice in this open room the early mine cars which were used to transport the broken rock out of the mine on rail tracks.

Returning to the mine entrance, note the old gold mill used in the early 1800's for extracting gold from the enclosing rock by means of the gravity stamp method, a process which pulverized the material. Looking directly south from the mine entrance, one can see workings of the Dakota Quartz Products (Division of Pacer Corporation) which is pulverizing quartz for sale to the Corning Glass Company for manufacture of glass products. The major mine workings in the hill to the south are those of the Hugo Mine, where another pegmatite is presently being worked primarily for quartz. This pegmatite is strictly off-limits for visitors. Looking southwesterly one can see the chairlift that provides a closer vantage point for viewing the faces in Mount Rushmore.

Stop 21.   Etta Pegmatite

Location.   This pegmatite is particularly well exposed at the Etta Mine approximately one mile south of Keystone (fig. 3.86).

Travel west out of Keystone to U.S. Highway 16A. Turn left (south) travel to a paved road at the location of the Whirlybird Rides helicopter pad and Rushmore Tramway. Follow this road for 0.2 mile to a small bridge and rock pinnacle at its intersection with a cross-road. Turn left onto the dirt road, travel 0.2 mile, and turn sharply left into the property of Dakota Quartz Products (Division of Pacer Corporation). At the Mill Office obtain permission from Mr. Donald Darrow, Manager, for entry to the Etta Mine. Then retrace the route to the road intersection at the rock pinnacle. Turn left and proceed up the gravel road, bearing right past the Keystone Cemetery sign, for 0.3 mile. At this point turn right and drive (or walk) along the poor gravel road, bearing left, for 0.3 mile to the entrance (just above an old wooden loading structure) to the Etta Mine tunnel (fig. 3.86). This tunnel was driven along the main mine level, at an elevation of 4585 feet, to intersect the central portion of the pegmatite. Exercise extreme caution within the mine.

History. Primarily because of its exceptionally large spodumene crystals and its world-wide position as the chief supplier of lithium for many years, this pegmatite is the most well known of the Black Hills pegmatites. Worked initially in 1881 for sheet mica, the deposit was mined subsequently for tin (cassiterite). The extraction of spodumene here in 1898 marked the inception of lithium mining in the United States (Norton et al. 1964). Spodumene mining was almost continuous until 1960 when the mine was closed. According to Norton et al. (1964), spodumene reserves are still present below the main tunnel level.

Description. In three dimensions the pegmatite has the shape of an inverted teardrop with its long axis inclined steeply northward Norton et al. 1964). In map view (fig. 3.90) it is crudely circular, slightly elongate north-south, highly irregular at the south end, and roughly 40,000 square feet in areal extent. Like the other commercial pegmatites in the Black Hills, the Etta Pegmatite is internally zoned. As shown in figure 3.90, six zones are defined on the basis of differing amounts of quartz, feldspars, micas, and spodumene in each unit. According to Norton et al. (1964), there may be at least one replacement unit (fig. 2.17) present in addition to the zones.

The pegmatite has intruded a quartz-mica schist representative of the medium-intensity, staurolite metamorphic zone (fig. 2.5). The schistosity of the rock is inclined steeply eastward; however, adjacent to the pegmatite, its orientation is typically the same as that of the contact. Around most of the pegmatite, the schist has been altered, in a zone 10 feet thick commonly, to a distinctive rock (granulite) which is massive and equigranular in texture. The granulite is composed of feldspars and quartz with varying amounts of micas and tourmaline (Norton et al. 1964).

Along the cut up to the tunnel entrance, the steeply inclined platy structure of the schist is well displayed. At the far end of the tunnel upon exiting from its southern spur (to the right), one crosses the contact with the pegmatite. As the contact is approached, the character of the schist changes dramatically across a three-foot-wide zone from a highly schistose rock to a relatively massive granulite rich in tiny black crystals of tourmaline (fig. 3.91). At the contact the pegmatite is represented by its well-developed, continuous wall zone

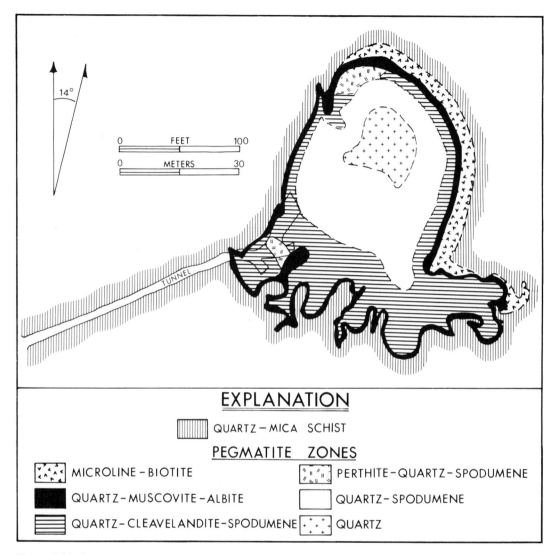

Figure 3.90. Generalized geologic map of the Etta Pegmatite (modified from Norton et al. 1964).

(fig. 2.17) labeled as the quartz-muscovite-albite zone on figure 3.90. Clearly visible here are muscovite, quartz, albite (sodic plagioclase) feldspar, and chalky gray crystals of spodumene. From this vantage point, looking directly east-northeast across the pit, one can see another contact between the pegmatite and schist. Although

now largely mined away, the pegmatite at this location was represented by its border zone (microcline-biotite zone on figure 3.90) which is present only along the northern and eastern periphery of the body.

Careful observation of the minerals exposed in the mine and comparison with figure 3.90 will enable one to detect readily

Figure 3.91. Contact (arrow) between the Etta Pegmatite (right) and schist converted to granulite (left).

most of the remaining four zones. For example, just northeast of the above location one can observe one of the two remaining (unmined) remnants of the perthite-quartz-spodumene intermediate zone (fig. 3.90). The dominant mineral in this zone is perthite (microcline containing thin lenses of albite) which occurs as crystals up to 15 feet long (Norton et al. 1964); a crystal four feet long is shown in figure 3.92.

Progressing from this location due east down into the pit to the east side, one can observe the quartz-cleavelandite-spodumene intermediate zone (fig. 3.90) which is particularly rich in spodumene (fig. 3.93). Widely exposed in the mine, this zone is characterized by a relatively continuous network of spodumene crystals attached to

one another and surrounded by sheaths of cleavelandite (platy albite) with the remaining spaces filled by quartz. To some, these relationships have suggested that the spodumene crystallized first as a self-supporting network surrounded by fluid from which cleavelandite then crystallized, followed by quartz (Jahns 1953). Others interpret this texture as representing the formation of spodumene from a lithium-rich fluid which replaced the previously crystallized minerals (Hess 1925).

From this location, looking west-northwest directly across the pit, one can observe the quartz-spodumene intermediate zone (fig. 3.90), the innermost of the three spodumene-rich intermediate zones. In this zone spodumene crystals up to tens of feet

Figure 3.92. Perthite-quartz-spodumene intermediate zone of the Etta Pegmatite, showing four-foot long crystal of perthite.

Figure 3.93. Quartz-cleavelandite-spodumene intermediate zone of the Etta Pegmatite, showing details of a large spodumene crystal.

Figure 3.94. Quartz-spodumene intermediate zone of the Etta Pegmatite, showing huge spodumene crystals. Note figure for scale.

in length (fig. 3.94), are surrounded largely by quartz. The quartz core zone of the pegmatite is represented largely by the island in the water-filled, lower part of the pit.

Well displayed is the glory-hole mining method, which here involved development of a vertical, cylindrical excavation (fig. 3.95). Access and ore removal were accomplished by tunnels at various levels.

Stop 22. Mount Rushmore National Memorial

Location. The Mount Rushmore Memorial is located approximately three miles southwest of Keystone. The rocks in the Memorial are well exposed in roadcuts

along Highway 244 which joins U.S. Highway 16A in the northeastern corner of the Memorial (fig. 3.96). In addition, the relationship between the two main kinds of rock in the area may be viewed in the distance from the Visitors' Center.

Description. Mount Rushmore itself consists of granite which has intruded a variety of mica-rich schists (fig. 3.97). Related to the granite is a third rock type, granitic pegmatite, which occurs as small dikes and sills scattered throughout the Memorial. As shown in figure 2.5, the granite here is a relatively narrow extension of the main mass of Harney Peak Granite to the west and south. Intrusion of the granite caused doming of the schists such that their platy structure, or schistosity, here, on the east

Figure 3.95. Aerial overview of the Etta Mine showing the glory-hole method of mining. Tunnel shown in figure 3.90 enters at arrow.

flank of the dome, is inclined easterly at a steep angle.

A starting point for observing the geology of the Memorial is the Visitors' Center. The View Terrace in the Visitors' Center was built in very coarse-grained granitic pegmatite similar to that present in other parts of the Memorial. Much of this rock is composed of large crystals of perthite (consisting of microcline containing thin lenses of albite feldspar) surrounded by somewhat smaller grains of quartz, albite (sodic plagioclase), and muscovite (Powell, Norton, and Adolphson 1973).

The view of the sculptures to the west (particularly the face of Washington) shows clearly the sharp contact between the Harney Peak Granite and the contorted schist below (See cross-section in figure 3.97). Below this major contact is a nearly horizontal dike of granite which cuts across the schistosity of the rocks it has intruded. In this area the granite is composed primarily of quartz and albite with smaller amounts of perthite and muscovite (Powell, Norton, and Adolphson 1973). A view of the sculptures shows the extensive jointing of the Harney Peak Granite and

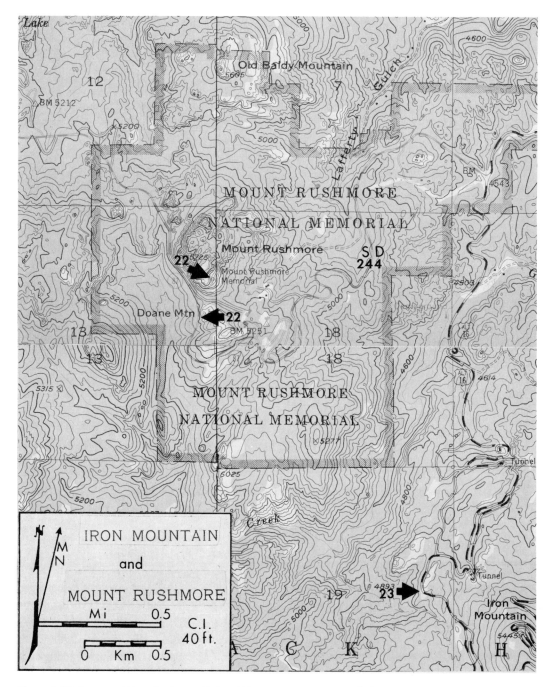

Figure 3.96. Stop 22, Mount Rushmore National Memorial.

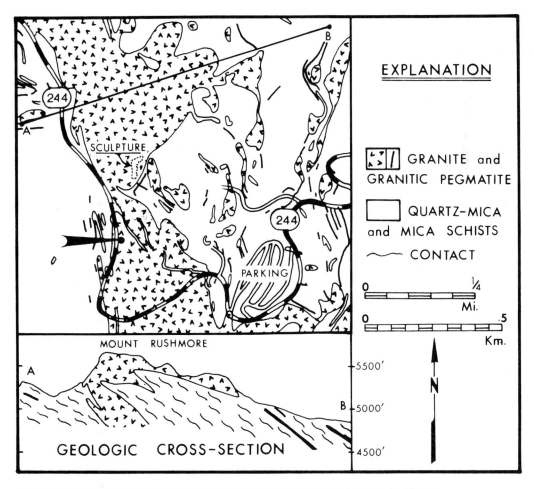

Figure 3.97. Generalized geologic map and cross section of the central part of the Mount Rushmore National Memorial (modified from Powell, Norton, and Adolphson 1973).

the great degree to which trees have rooted in the joints and in the rubble below the faces. The highest point of the sculptures is at an elevation of 5,659 feet, roughly 400 feet above the View Terrace which is slightly less than one mile above sea level.

Almost due south of the sculptures on the east side of Highway 244 (fig. 3.96), one can observe the schists and related rocks which were intruded by the Harney Peak Granite. Along the roadcut note that the schistosity generally is inclined steeply eastward, consistent with the position of these rocks on the east flank of the dome. However, locally this structure has been deflected in proximity to forcefully intruded dikes. A good example of this exists at the south end of the roadcut where the schistosity makes a right-angle bend adjacent to an eight-foot-thick granitic pegmatite body (fig. 3.98). Near the contacts this particular pegmatite contains scattered small purple garnets. Local zones of in-

Figure 3.98. Deflection of the schistosity in rocks intruded by a granitic pegmatite body.

tense yellow and red reflect weathering of the schists related to their iron-bearing minerals which are undergoing alteration to iron oxides. A somewhat thicker pegmatite occurs about half-way northward along the roadcut. Notice that the pegmatitic granite thoroughly impregnates the schist which is intensely deformed in the areas near these thin bodies of granitic material. About two-thirds of the way north along the roadcut there are a large number of ellipsoidal structures in a coarse-grained hornblende-plagioclase schist (fig. 3.99). These structures are flattened calc-silicate nodules derived by the metamorphism of calcareous concretions in the original sedimentary rock. At the north end of the cut there is a large amount of granite and pegmatite in a zone of schists which are intensely deformed into very tight, angular,

chevron folds (fig. 3.100) as well as more gently curving folds. At least one of these pegmatites, a lenticular pod about six feet long by three feet wide, shows well-developed textural zoning from a moderately coarse rim, about one foot thick, to a core containing feldspar crystals and quartz masses up to one foot in maximum dimension (fig. 3.101). The pegmatite is composed primarily of perthitic microcline, quartz, muscovite, and biotite. The very coarse-grained core of this body exhibits perfectly formed crystals of perthitic microcline intergrowth with granular, smokey gray quartz. Looking north from the south end of this roadcut, one can see the sculpture of George Washington's face and, to the left, excellent weathered joints which are typical of the Harney Peak Granite throughout this area (fig. 3.102).

Figure 3.99. Calc-silicate nodules formed by metamorphism of calcareous concretions.

Figure 3.100. Chevron folds exhibited by biotite schist and associated layers.

Figure 3.101. Well-developed textural zoning in a small granitic pegmatite lens.

Figure 3.102. Vertical joints in Harney Peak Granite. Note sculpture of Washington's face in background to right.

Most of the metamorphic rocks in the Memorial were originally shales or impure sandstones which have been converted to mica schist and quartz-mica schist. Microscopic studies of these rocks indicate that they contain typically 15-30% biotite, 10-30% muscovite, and 30-80% quartz, as well as small amounts of other minerals including sillimanite (Powell, Norton, and Adolphson 1973). The latter mineral establishes these rocks lie within the sillimanite zone, the zone which reflects the greatest intensity of metamorphism in the Black Hills (fig. 2.5).

Sculpture.   The faces of the four American presidents were created at Mount Rushmore by Gutzon Borglum, a sculptor who had earlier fashioned the Confederate memorial at Stone Mountain, Georgia. In 1924 Borglum was invited by the South Dakota state historian to visit the Black Hills and to consider a proposal of his to create figures of several western heroes in the Needles area. Borglum suggested the presidents as subjects and ultimately selected Mount Rushmore for the site because of the uniform grain size of the granite there, the fact that the mountain dominated the surrounding terrain, and because the mountain faced the sun most of the day. In 1927 Mount Rushmore was officially dedicated as a national memorial, and work was initiated on the project. Although actual construction time amounted to 6.5 years, the work was not completed until 1941, 14 years later, due largely to inadequate funding and poor weather. The initial funding for the project was finally supplemented by federal fund-slightly under one million dollars.

Figure 3.103 shows the Mount Rushmore site as it looked in 1924 when Borglum picked it for his work. His first job was to select a grouping of presidential faces which would fit best on the exposed granite face and which could accommodate the extensive jointing characteristic of the rock. In order to guide the workmen, he constructed models, each measuring five feet from chin to top of the head. The scale selected was such that one inch on the model was equivalent to 12 inches on the mountain. Critical measurements for each model were made with a plumb bob from a horizontal bar and then transferred to the mountain by means of a 30-foot-long movable boom. After each reference point, such as the tip of the nose or corners of each eye, was located on the mountain, the excess rock was removed by use of dynamite. During this phase of the work, some 450,000 tons of rock were removed.

As figures 3.104 and 3.105 show, the work progressed from a complex system of scaffolding suspended along the side of the mountain. In addition, individual workers were supported by "swing seats" which they could control by hand-operated winches. From these seats they used jackhammers to drill holes into which the dynamite was inserted. Blasting served to remove the rock to within three or four inches of the final surface. At this stage, holes were drilled at roughly three-inch intervals over the surface. The resulting network of holes facilitated removal of the remaining rock by means of small drills as well as hammers and wedging tools. A final smooth finish was obtained with a small air hammer utilized in a process referred to as "bumping." The resulting faces (fig. 3.106) were developed at a scale consistent with figures 465 feet tall. Typically each of the heads measures 60 feet from chin to

Figure 3.103. Mount Rushmore site as it looked prior to 1927 when work was begun on the sculptures (photo provided by the National Park Service).

Figure 3.104. Overview of construction work on the sculptures. Note sharp contact between granite and schist below (photo provided by the National Park Service).

Figure 3.105. Closeup showing construction work on the Lincoln sculpture (photo provided by the National Park Service).

Figure 3.106. The Mount Rushmore sculptures today. From left to right, Washington, Jefferson, Roosevelt, Lincoln (photo provided by the National Park Service).

top. Each nose is approximately 20 feet long, each mouth 18 feet wide, and each eye about 11 feet across.

Stop 23. Iron Mountain Pegmatite

Location. This pegmatite is intersected by U.S. Highway 16A west of Iron Mountain, approximately 0.3 mile south of a tunnel on highway (fig. 3.107). The main portion of the pegmatite has been exposed on the east side of the road cut and may be traced up the hillside east of the road. Parking for one or two cars is available via a small turnoff just south of the pegmatite.

Description. This small stock-like (fig. 2.13) pegmatite body, well exposed in the road cut, is generally representative of a number of the larger commercial pegmatites in the Black Hills region. Careful examination reveals the presence of crude zoning (fig. 2.17) with a rose quartz core. Particularly in the road cut, one can observe the extremely coarse-grained texture of this rock which is dominated by large crystals of microcline feldspar, four feet or more in maximum dimension, and possessing unusually well-formed crystal faces (fig. 3.108) with which white quartz and a large mass of rose quartz are associated. Close observation of the microcline reveals that it contains thin lenses or films of a translucent mineral within it. This mineral is plagioclase feldspar, and its presence defines the texture of the microcline as "perthitic" (an intergrowth of plagioclase with microcline). Note the strong development of striations (fig. 3.109) in some of the rose quartz; these are reflections of interruptions during crystal growth. Careful

examination of the rose quartz near road level reveals the presence of a large (4-inches across), well-formed, six-sided (hexagonal) crystal of pale, green beryl (fig. 3.110) which has been largely removed, leaving a deep six-sided cavity with beryl at its base.

Progressing northward toward the curve in the road one can observe a reduction in the grain size of this pegmatite as the contact is approached. In the contact area the other more common pegmatite minerals are intergrown with a large amount of jet-black tourmaline crystals which are typically less than one-half inch in diameter.

Away from the road to the east, the pegmatite grades into more typical, variably textured Harney Peak Granite. Several large inclusions of schist occur in this rock, particularly near the top of the hill.

Looking northwestward toward the Mount Rushmore Memorial which is visible in the background, one gets an excellent view of Mount Baldy (fig. 3.111), a feature typical of the topography which has developed on the well-jointed and fractured Harney Peak Granite.

Stop 24. The Needles

Location. The major features of the Harney Peak Granite can be observed by traveling west from Keystone on U.S. Highway 87 through the Mount Rushmore National Memorial and then eventually south through the Sylvan Lake area and on Highway 87 along the Needles Highway through The Needles area (fig. 3.112).

Description. Along this route, particularly upon approaching the Needle Eye, one

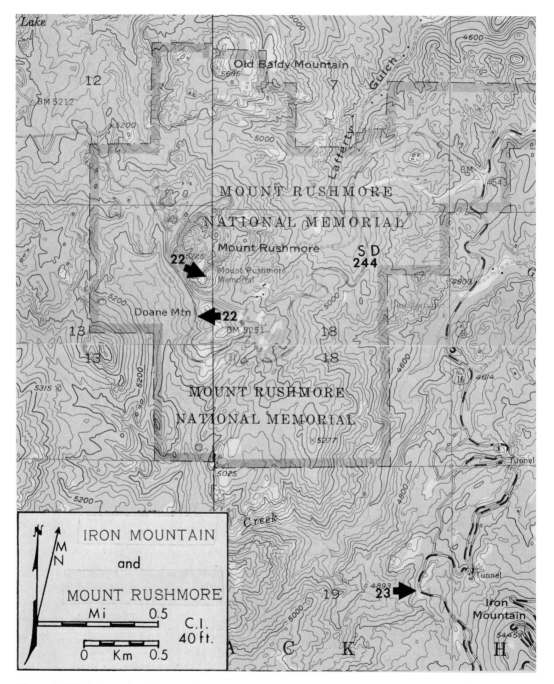

Figure 3.107. Stop 23, Iron Mountain Pegmatite.

Figure 3.108. Huge, well-formed microcline crystal.

Figure 3.109. Growth striations in rose quartz.

Figure 3.110. Large six-sided crystal of beryl in rose quartz.

Figure 3.111. View, northwestward from the Iron Mountain Pegmatite, of Mount Baldy showing strong jointing and irregular fractures in the Harney Peak Granite. Note Mount Rushmore sculptures (arrow).

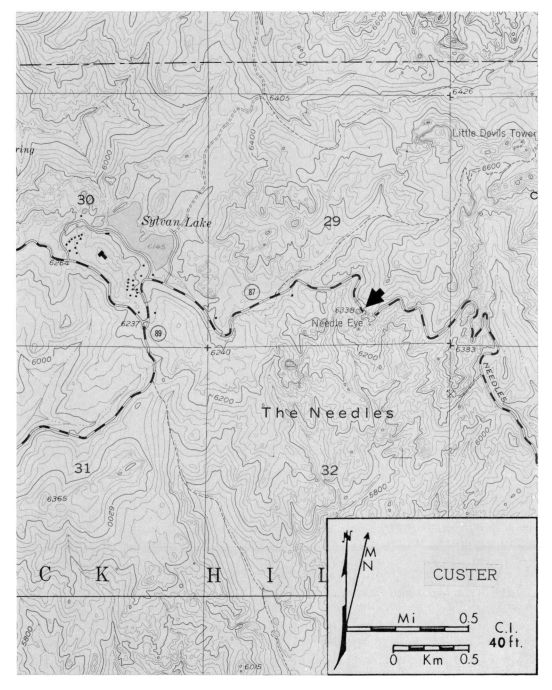

Figure 3.112. Stop 24, The Needles.

Figure 3.113. "Needle-like" topography developed on well-jointed Harney Peak Granite.

can see the effect of erosion of the well-jointed Harney Peak Granite. The joints vary in their spacing, and erosion along them has produced spectacular topography characterized by jagged pinnacles of granite separated by deep slots from which the granite has been removed (fig. 3.113). The jointing is particularly well displayed at the Needle Eye where weathering along the joint system has produced a lenticular slot which suggests the eye in a sewing needle (fig. 3.114).

At the Needle Eye locality, one can observe the great variability in grain size displayed by the Harney Peak Granite. The granite varies between true granite, of medium to coarse grain, to granitic pegmatite consisting of very large crystals. As shown in figure 3.115, the grain size variations in the granite suggest a sub-horizontal layering. Just east of the Needle Eye, on the north side of the tunnel developed in a

large joint (fig. 3.116), one can see an inclusion of schist within the granite. This mass is a remnant of the host rock into which the Harney Peak Granite was intruded.

Stop 25.   Jewel Cave

Location.   Jewel Cave National Monument about 12 miles west of Custer, South Dakota on U.S. Highway 16 (fig. 3.117). Access to the cave is by National Park Service tours.

Description.   Jewel Cave is one of a number of caves developed in the Black Hills region; a description of it can serve as a general description of caves and cave formation in the Black Hills. All were formed in the Pahasapa Formation and they differ from one another only in size and the de-

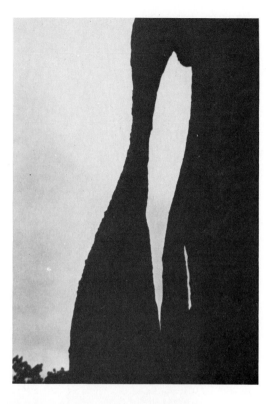

Figure 3.114. The Needle Eye.

Figure 3.115. Subhorizontal layering in the Harney Peak Granite defined by grain-size variations at the Needle Eye.

Figure 3.116. Tunnel along U.S. Highway 87 developed in a partially eroded joint within the Harney Peak Granite at the Needle Eye.

gree to which cave deposits are developed. Jewel Cave is the largest cave in the Black Hills and, indeed, one of the largest in the world. Well over 40 miles of passageways have been explored and mapped in this cave.

The formation of Jewel Cave, as with any cave, is a result of the interaction of four geologic conditions. First, suitable rocks must be available for the formation of cave passageways. Second, these rocks are typically fractured to provide selective passageways for the movement of groundwater. Third, the rock must be located in such a position relative to the groundwater table to afford alternate periods of solution and deposition of material. Fourth, the chemical composition of the water must be conducive to solution and deposition of soluble material.

In the Black Hills the Pahasapa Formation constitutes a large amount of soluble rock in which cave development can occur. The rock was originally formed in the Mississippian Period by accumulation of fragments of calcium carbonate in a shallow marine environment. The bulk of the fragments making up the unit consisted of broken remains of marine organisms and calcareous mud which accumulated to a thickness of about 400 feet in the Black Hills region. As this rock was buried by subsequent deposits, the individual grains were tightly bound together by calcium carbonate resulting in the rock limestone.

During the Laramide Orogeny, the Pahasapa Formation and the units above and below it were uplifted more or less to their present position. One of the results of this uplift was that a system of joints, or fractures, developed in most of the rock units. In the area of Jewel Cave the orientation of these joints formed a cross-hatched pattern with one set of joints oriented essentially north-south and the other east-west. These joints formed two of the three planes along which groundwater could move more easily than it could pass through the pore spaces in the limestone. The third plane of transport of water was parallel to the beds or layers that had originally formed at the time that the Pahasapa Formation was deposited. These planes dip gently to the west, consistent with the regional dip of beds in the Jewel Cave area.

Early stages of development of the cave occurred during a time when the Pahasapa Formation was below the groundwater level. At this time, water moved along the joint planes and along the bedding planes flowing slowly down the dip of the beds in a westerly direction. This water dissolved some of the limestone from the Pahasapa Formation enlarging the fractures

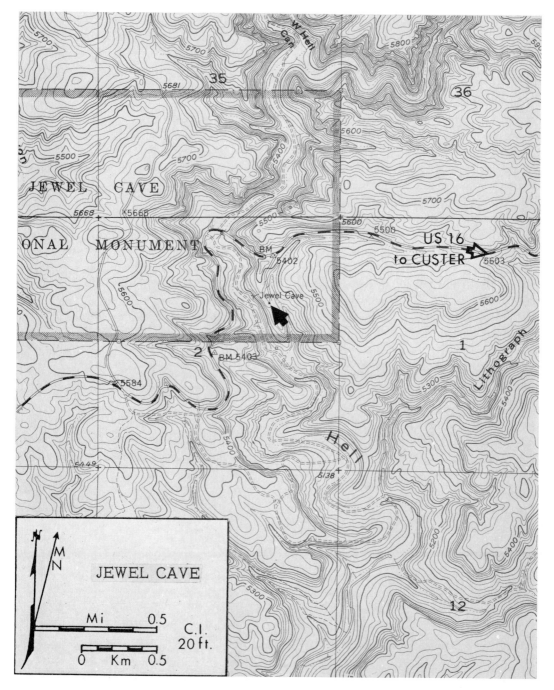

Figure 3.117. Stop 25, Jewel Cave National Monument.

and the bedding plane surfaces to form an intricate network of passageways.

Subsequent to the solution of these passageways, the chemistry of the groundwater changed. Whereas the water had originally dissolved calcium carbonate from the Pahasapa Formation, conditions changed to permit precipitation of calcium carbonate. Although this could have resulted either from change in the concentration of calcium carbonate in the groundwater or change in temperature of the water, it seems most likely that temperature change was the factor contributing to precipitation of material in Jewel Cave. Because groundwater flowing through the Pahasapa Formation would always be heavily charged with calcium carbonate, there would never seem to be a lack of it in the water. However, by reducing the water temperature,

calcium carbonate becomes much more soluable and passageways could form by speeding the rate of solution. Conversely, warming of the groundwater would result in deposition of calcium carbonate. A change of groundwater temperature was probably associated with general climatic change because a close relationship exists between mean, annual temperature of an area and the temperature of the groundwater.

Precipitation of the mineral calcite can take many different forms in a cave environment; nearly all varieties of cave deposits are visible in Jewel Cave. These deposits can be arranged in two general categories. Crystalline deposits of calcite typically form under water by slow precipitation of material around centers of growth (fig. 3.118). Most of the passage-

Figure 3.118. Crystalline calcite, dogtooth spar, deposited in Jewel Cave at a time when the cave was flooded. Photo courtesy of the National Park Service.

ways in Jewel Cave are lined with this kind of crystalline material, termed dogtooth spar or nailhead spar depending upon the bluntness of the elongate crystals formed. Growth of crystals is clear indication of precipitation at a time when the area was below the groundwater level. Perhaps the more common kind of cave deposits are features collectively referred to as dripstone and flowstone—stalactites and stalagmites, columns, draperies, etc. (fig. 3.119 and 3.120). These deposits are composed of calcareous material, referred to as travertine, which typically form above the groundwater level. They are produced when calcium-rich water flows or drips down into an opening or cave passageway and slowly evaporates, leaving the calcium carbonate as a deposit. Dripstone and flowstone do not form below the water table but do require the passage of water heavily charged with calcium carbonate.

Although calcium carbonate is the dominant mineral in formations within Jewel Cave, other minor formations consist of siliceous material, manganese dioxide, and gypsum. These latter minerals form incrustations on passageways otherwise formed of calcium carbonate deposits.

At the present time active mineral deposition is very limited within Jewel Cave. Although groundwater continues to move through the rocks in the Black Hills region, the rate of accumulation of material within the cave is relatively slow. However, this could change with a change in water chemistry or elevation of the groundwater table. Evidence that such events have occurred in the past consists of the presence of clay-like layers incorporated in cave deposits. This is suggestive of former alternating conditions of elevation above and subsidence below the water table. Therefore, although we see the cave today as "finished," it re-

Figure 3.119. Columns, stalactites, and draperies formed of dripstone in Jewel Cave. Photo courtesy of the National Park Service.

mains a dynamic feature which will continue to be modified as conditions in the Black Hills region change with time.

Stop 26.   Custer Schist

Location.   This stop is a road cut just east of downtown Custer at the intersection of U.S. Highway 16A with Highway 89 (fig. 3.121).

Description.   At this locality one can observe rocks representative of the schists in the southern part of the Black Hills. The rocks exposed in the cut (fig. 3.122) constitute part of a thin quartz-biotite-garnet

Figure 3.120. Flowstone, locally referred to as "popcorn," which covers many cave areas in Jewel Cave. This material may have formed by deposition of calcium carbonate from water that splashed or dripped onto the surface. Photo courtesy of the National Park Service.

schist unit mapped by Redden (1968) to the northwest of Custer and southward for almost 15 miles to Pringle. In the Custer area much of the schist is coarser grained than that in the northern part of the Black Hills; and many of the varieties of schist here contain sillimanite, indicating that these rocks belong to the highest-intensity sillimanite metamorphic zone (fig. 2.5).

In the road cut the rocks are largely interlayered quartz-rich and mica-rich schists which locally contain small purple garnets. Within the steeply inclined foliation planes (schistosity) are elongate minerals and mineral aggregates which define a steeply plunging lineation (fig. 3.123); planar and linear structures are common in the schists of the Custer area. Figure 3.124 is a microscope view of a representative sample of the schist collected from the midpoint of the road cut (fig. 3.122). As shown, the rock is dominated by quartz with which biotite and muscovite micas occur as well as tourmaline and other minor minerals.

Portions of these rocks are cut by quartz veins, a common feature of this area. According to Redden (1963, 1968) who has studied many such veins in two large areas west and southwest of Custer, these features range in thickness from fractions of an inch to more than 100 feet. Typically a single vein contains less than 5% of minerals

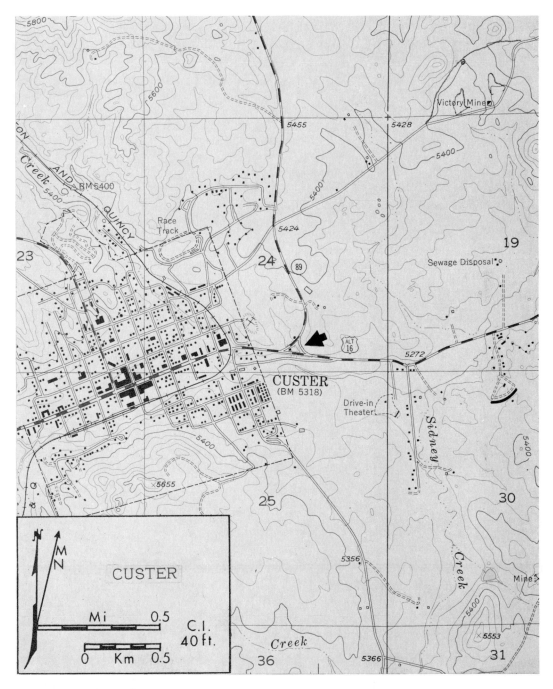

Figure 3.121. Stop 26, Custer schist.

Figure 3.122. Schist overlain by sand and gravel at intersection of U.S. Highways 16A and 89.

Figure 3.123. Closeup of schist showing steeply inclined foliation and lineation.

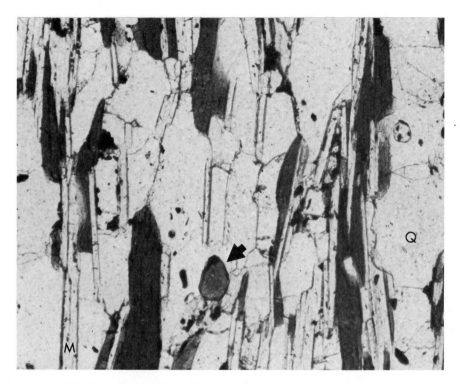

Figure 3.124. Photomicrograph of schist showing elongate, aligned grains of quartz (Q), biotite (dark grains), and muscovite (M) as well as one weakly-zoned tourmaline crystal (arrow). One nicol, 100X.

other than quartz. In many of the veins these minerals include common metamorphic minerals (sillimanite, staurolite, garnet) which also occur in the adjacent rock. In other veins native gold, or wolframite (a tungsten mineral), is an associate of the quartz, and the adjacent schist is characteristically altered along each vein (Redden 1968). Redden (1968) suggested that the former veins may represent material redistributed during metamorphism, whereas the latter may have been introduced to the schists during the interval of igneous activity in which the granitic pegmatites formed.

In the upper part of much of the road cut, one can observe that the schist has been deformed into highly angular chevron folds (fig. 3.122). Such structures are the result of relatively recent surface creep due to gravity effects. Overlying the schist is a deposit of sand and gravel which, at the north end of the cut, appears to occupy a channel eroded into the schist (fig. 3.122). Gold was first discovered in the Black Hills in these gravels.

Stop 27. Bull Moose Pegmatite

Location. Access to this locality is provided by traveling east of downtown Custer on U.S. Highway 16A, just past its intersection with Highway 89, and turning sharply right (south) onto a partially paved road (fig. 3.125). Follow this road

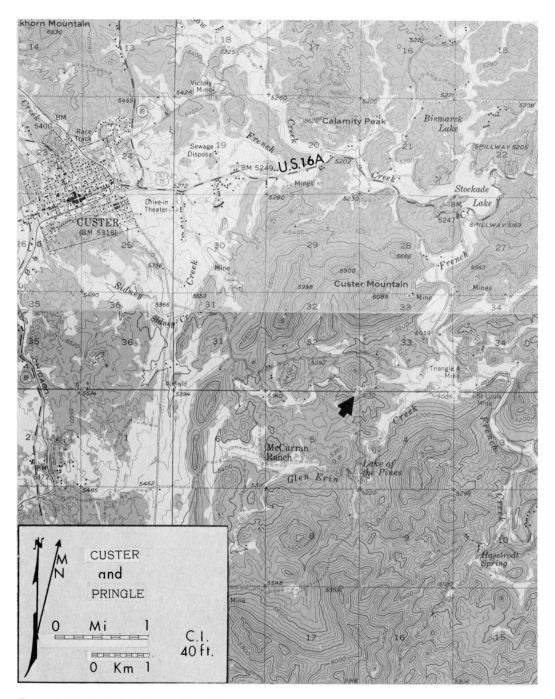

Figure 3.125. Stop 27, Bull Moose Pegmatite.

for 1.8 miles, bearing left at the fork, and make a sharp left (east) at the sign to French Creek on County Road 343. Travel 2.6 miles along Road 343 and turn right (south) onto a gravel road. For 0.4 mile, follow the gravel road to a cluster of stucco and wood houses. Permission to enter the mine must be obtained here from Mr. James Bland. The mine is another 0.2 mile along the dirt road which crosses in front of the houses.

Description. The Bull Moose Pegmatite is representative of the zoned granitic pegmatites in the southern Black Hills area (Norton 1964). As shown in figure 2.16 the area around Custer, and between Custer and Pringle to the south, is a major pegmatite district in terms of the number of such bodies per square mile.

As one approaches the Bull Moose Mine, the small, relatively new tunnel on the right exposes the metamorphic rock into which the pegmatite was intruded (fig. 3.126). The dominant rocks are interlayered quartzite and quartz-mica schist. The schist is relatively coarse grained and dominated by quartz, biotite, muscovite, and sillimanite (fig. 3.127). Both features are consistent with the fact that these rocks lie within the highest-intensity sillimanite metamorphic zone (fig. 2.5).

By following the road beyond the new tunnel, one reaches the main portion of the mine which can be entered by means of an inclined road. Looking straight ahead from the entrance, one can see the roof of the pegmatite in contact with overlying schist as well as in inclusion of the schist (fig. 3.128). Careful examination of the

Figure 3.126. Interlayered quartzite (massive beds) and quartz-mica schist (platy zones) near the Bull Moose Pegmatite.

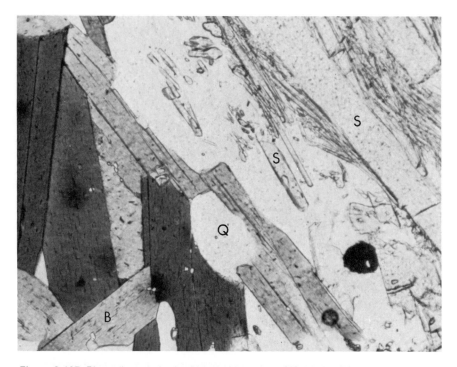

Figure 3.127. Photomicrograph of schist showing quartz (Q), biotite (B), and sillimanite needles and prisms (S). One nicol, 100X.

Figure 3.128. Bull Moose Pegmatite showing schist (S) inclusion (right center) and contact with schist near top of pegmatite (arrow).

Figure 3.129. Part of a large crystal of muscovite showing prominent cleavage (planar surfaces).

pegmatite along the inclined road reveals the presence of a crude zonation inward from a moderately coarse-grained area rich in muscovite and perthite (microcline containing thin lenses of albite), to an area of coarser-grained pegmatite rich in perthite, to an extremely coarse-grained core rich in muscovite and rose quartz. Rose quartz masses attain a maximum dimension of six feet in the core. Some of the muscovite crystals are up to one-foot across (fig. 3.129); and jet black tourmaline crystals, common in this pegmatite, may be one or two inches in diameter (fig. 3.130). Also relatively common are brownish-black crystals of columbite-tantalite, a rare mineral which constitutes a commercial source of niobium and tantalum.

Stop 28. Buffalo Gap

Location. Custer County, South Dakota. Access to this locality is via Custer County Highway 101. From the junction of U.S. Highway 18 and South Dakota Highway 79, east of Hot Springs, travel approximately 7.5 miles north and turn left (west) on Highway 101. The locality is about three miles west of this turn (fig. 3.131).

Description. Rocks exposed at this locality show an exceptional example of ancient stream deposits that have been affected by hot spring activity. The rocks that form Buffalo Gap include the red shale of the Triassic Spearfish Formation, the Jurassic Sundance Formation, and the Jurassic

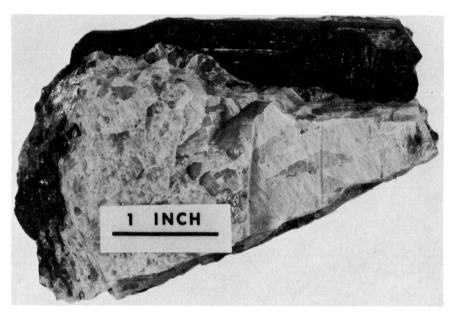

1 INCH

Figure 3.130. Large dark tourmaline crystal.

Unkpapa Formation. All of these rocks can be observed from road level by viewing the hillside north of the highway. However, access to the bluff itself is across private land, and permission must be obtained to examine the rocks in more detail. Adjacent to the road, however, is an 18-foot-thick deposit of boulders and cobbles (fig. 3.132) which are remnants of a stream deposit much younger than the surrounding rocks. Although not dated precisely they are thought to be of Pleistocene age and to represent deposits formed during an earlier stage of the development of Beaver Creek. Examination of the boulder conglomerate reveals that it is very well cemented by calcium carbonate and that the base of the conglomerate is cut into the Spearfish Formation (fig. 3.133). The color contrast between the red shale of the Spearfish and the lighter colors of the conglomerate make recognition of this contact straightforward. Also cutting through the Spearfish Formation are several dikes, or fracture-fillings, of calcium carbonate (fig. 3.133). It appears that this material has been deposited in fractures in the Spearfish Formation by passage of groundwater upward into the conglomerate and has acted as a source of the cement that binds the boulders together. The high concentration of calcium carbonate, which acts as the cementing agent, was probably derived from warm groundwater issuing out at the surface as a hot spring. Hot springs are common in this part of the Black Hills; and they can be observed today at Cascade Springs (Stop 30) and in the city of Hot Springs. Although no hot springs are currently active in the Buffalo Gap area, they certainly must have existed during the time that the conglomerate was cemented.

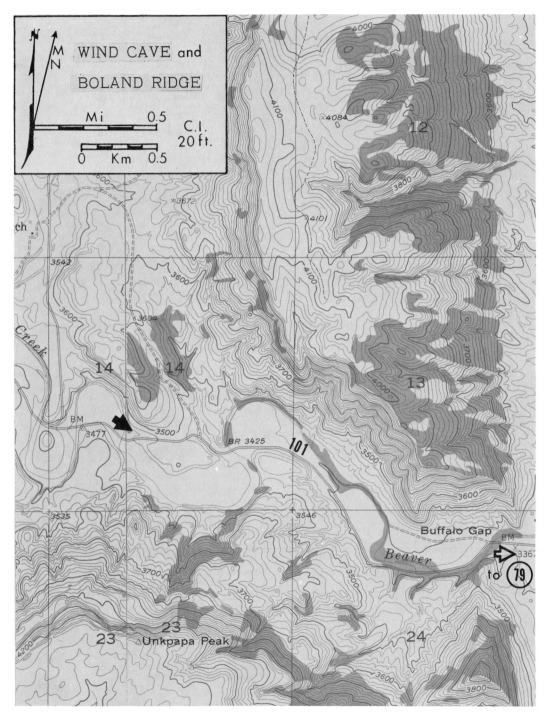

Figure 3.131. Stop 28, Buffalo Gap.

Figure 3.132. Pleistocene stream gravel in an ancient channel of Beaver Creek. This channel deposit closely resembles those found in fast moving streams in the Black Hills today but is located almost 20 feet above the active stream and is firmly cemented by calcium carbonate.

Figure 3.133. Sharp contact between a Pleistocene channel and the Spearfish Formation. The steeply dipping feature is a "dike" of calcium carbonate following a joint in the Spearfish.

Another extensive deposit of this type of conglomerate can be observed along the valley of the Fall River in downtown Hot Springs. However, at this locality none of the feeder dikes of calcium carbonate are exposed, so that observations related to the activity of groundwater passage and cementation of the boulders is less clear.

Stop 29. Hot Springs Mammoth Site

Location. Hot Springs, South Dakota. From Highway 385 turn west on University Avenue and cross the viaduct. Travel to 19th Street and turn left (south) on 19th and proceed to Evanston Avenue. The site is one block east on Evanston (fig. 3.134).

Description. In 1974 excavation for a housing development at this locality uncovered bones of Pleistocene mammoths (extinct elephants) in a sequence of silts and clays. Subsequent excavation has exposed an unusually large number of bones of these animals along with the remains of contemporary organisms. Under the supervision of paleontologists from Chadron State College, Chadron, Nebraska, some of the specimens have been removed from the site, but most have been retained in their original positions for observation by the public (fig. 3.135).

Although the location now occurs on a hillside and the top of a hill, it appears that at the time the animals were preserved in the sediment the site was a depression, probably a small pond occupying a sink hole, or solution depression, similar to those which exist in the area today (fig. 3.136). The rocks in the immediate vicinity of the site are part of the Spearfish Formation of Triassic age. Directly beneath the Spear-

fish is the Minnekahta Formation, the source of many springs, one of which is the namesake for the city. A similar spring may very well have occupied the mammoth site during Pleistocene time with water from the Minnekahta Formation coming to the surface in a small depression developed in the Spearfish. Accumulation of silt and clay provided the burial site for the fossil material. Thus far not only have mammoths been collected, but also bears, camels, peccaries, coyotes, and a bird, along with freshwater clams and snails.

Detailed examination of the remains of the organisms can yield considerable information about how they died and were preserved. For example, enough mammoths have been collected and studied to suggest the following scenario (Laury 1978). Animals approaching the pond to drink may have occasionally encountered unstable or slippery footing and thus were trapped in the sediments at the bottom of the sinkhole. Most of the remains occur around the margins of the sink; very few are found near the center of the deposit. Upon entrapment, some of the animals were buried almost immediately and, therefore, the bones are articulated, i.e., arranged in life position. Others apparently were not buried until they had been preyed upon by other animals, decomposed, and disarticulated. In either event, the preservation of the material was extremely good and is suggestive of rather rapid burial. Examination of one element of the mammoth skeletons, notably the mandibles or lower jaws, suggests that at least fifteen individuals are buried at this site. An equal number may yet be discovered here.

Subsequent to preservation of the animals and filling in of the sinkhole with sediments, the surrounding Spearfish sedi-

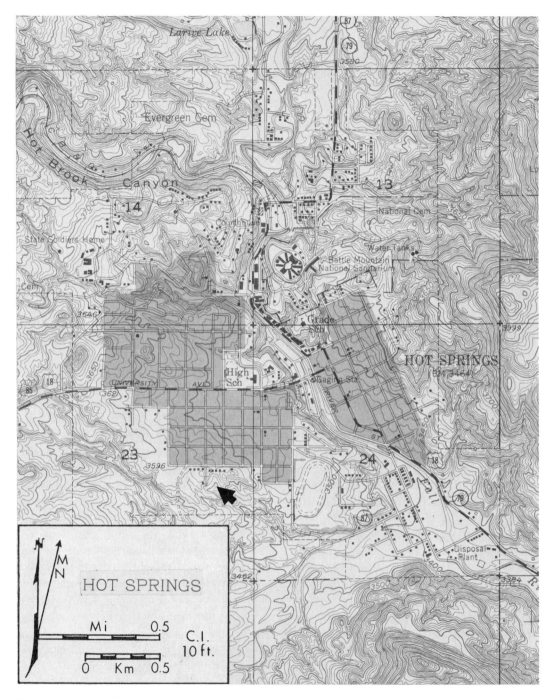

Figure 3.134. Stop 29, mammoth site, Hot Springs, South Dakota.

Figure 3.135. Mammoth skull with tusks exposed at the mammoth site and preserved at the locality. Note that two additional tusks are visible to the left of the skull.

Figure 3.136. Sink hole developed in thinly-bedded Minnekahta Limestone east of the mammoth site. Note the trees which are rooted in soil at the bottom of the sink hole, well below ground level of the field.

ments were eroded away, leaving the deposit perched on the hillside. This discovery preserves an outstanding assemblage of organisms that inhabited the Black Hills region during Pleistocene time. In fact radiometric dating, using the Carbon-14 technique, suggests that the age of the site is about 20,000 years (Agenbroad 1977). Although the age of the sequence is not particularly unusual relative to the occurrence of the mammoths and associated animals, there is one aspect that is potentially even more exciting. An artifact has been tentatively identified from the site. A small stone fragment, found in association with the mammoths, may be part of a scraping tool used by some of the first human inhabitants of this region. Although not yet confirmed, it introduces the tantalizing possibility that man was in the Black Hills region as early as 20,000 years before current times (Agenbroad 1977). Much more work will need to be done, however, before this can be substantiated.

Collection and preparation of the material from the fossil site involves detailed, painstaking work. The sediment is carefully removed using small digging instruments and brushes to clean the bedding surfaces (fig. 3.137). When bone material is exposed, it is tentatively identified, its position is mapped both vertically and horizontally, and it is impregnated to harden it and protect it from weathering. If the bone is to be removed from the site, it must be even further protected. In this case, after much of the sediment has been removed from around the bone material it is sheathed in plaster of Paris (fig. 3.138) to protect against breakage and finally removed. Once the specimen arrives at the laboratory, its protective plaster encasement is removed, and the specimen is further cleaned, prepared, and finally studied.

Stop 30. Cascade Springs

Location.   About eight miles south of Hot Springs, South Dakota. Access to the locality is via South Dakota Highway 71. From the junction of U.S. 18 (Truck Bypass) and South Dakota 71, travel 8.3 miles south on 71 to the R.H. Keith Park. Park and walk to the springs located just east of the road (fig. 3.139).

Description.   Underground water emerges at the surface as springs in many places throughout the Black Hills. In most instances the springs can be related to the passage of water through Paleozoic limestone and dolostone. Recent studies (Rahn and Gries 1973) have classified the springs in the Black Hills region into six categories, of which three are considered major occurrences. The types are defined primarily on the stratigraphic occurrence of the springs such that one type is confined to carbonate, water-bearing rocks (aquifers) where water enters the carbonate rocks from the surface, moves through it, and emerges at some other locality. Another type is characterized by carbonate rocks charged with water which passes down through the rocks and emerges along their base, or lower contact. The third type, such as Cascade Springs, includes occurrences of springs along the contact between the Spearfish Formation and the underlying Minnekahta Formation. Of the three types the springs along the Minnekahta-Spearfish contact tend to be the most persistent in that they do not show a great deal of seasonal fluc-

Figure 3.137. Sediment is being removed carefully to search for additional fossil material. Note the fine, inclined layering of the lake sediments.

Figure 3.138. Preparation of a specimen for removal requires covering the bone with a "cast" of plaster of Paris. This protective coating can be removed in the laboratory allowing the material to be studied.

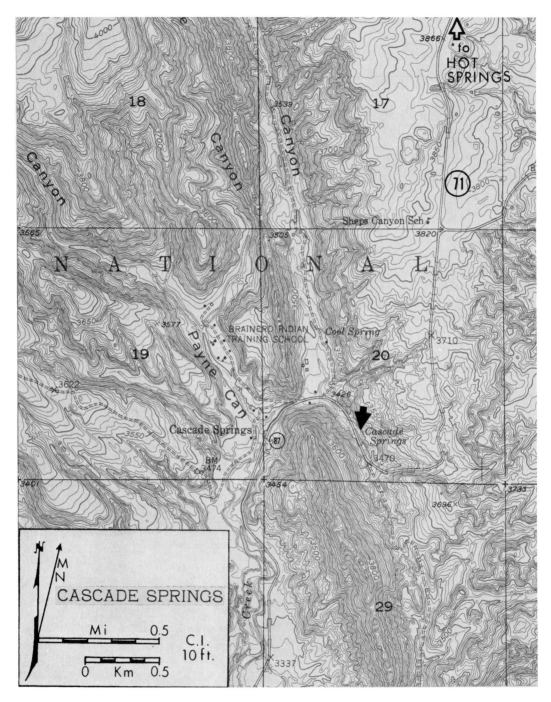

Figure 3.139. Stop 30, Cascade Springs.

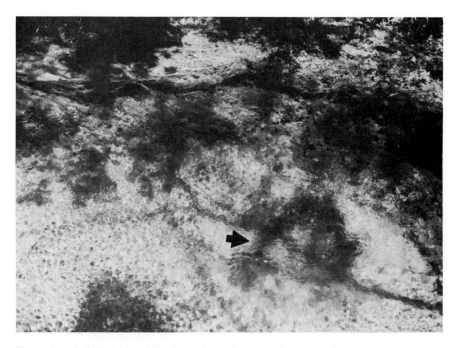

Figure 3.140. Orifice in the Minnekahta Formation through which water emerges onto the surface. Several such openings occur in the Cascade Springs area.

tuation. The others are more directly related to surface discharge and rainfall and tend to fluctuate seasonally. At Cascade Springs one can observe water bubbling up from the top of the Minnekahta Limestone through a large (18-inch diameter) solution cavity in the Minnekahta (fig. 3.140). Just above the springs the Spearfish Formation can be identified (fig. 3.141). It is bright red, and, although it is weathered deeply in the immediate vicinity of the springs, it is a red, gypsum-bearing shale on the adjacent hillside.

The movement of groundwater through the carbonate rock units in the Black Hills suggests a relationship between the origin of these springs and the much larger cavities and caverns that produce the numerous caves throughout the Black Hills. The limestone units tend to be strongly jointed and water passing along these fractured surfaces has dissolved the soluble carbonate minerals producing open passageways. In many places these passages have grown to such size that large caves have developed. More commonly, however, the openings are relatively small but certainly large enough to provide passageways for the movement of groundwater. Water that moves through these systems has its origin as surface water which has percolated into the rocks either through cracks and fractures in the rocks or through the pore spaces between individual grains that make up those rocks. The rate of flow is clearly related to the size of the openings as well as to the gradient, or slope, along which the water is moving. In the case of Cascade Springs, a

Figure 3.141. Contact between the Minnekahta Formation (below arrow) and the Spearfish Formation (above). This horizon is the most common zone of persistent spring activity in the Black Hills.

relatively consistent flow of about 23 cubic feet of water per second has been recorded, making it one of the largest springs in the Black Hills. The water emitted from springs in the area, although plentiful, is rather heavily mineral-charged as a result of solution of the carbonate aquifers.

As might be expected, most of the water issuing from springs in the Black Hills is relatively cool. In fact, the water temperature of most springs varies from about 44° to 47°F., which corresponds to the mean annual temperature of the area. In the southeastern part of the Black Hills, however, a number of springs have abnormally high temperatures. This zone, centered at

Hot Springs, South Dakota, produces water varying in temperatures from 60°F. to as much as 87°F. at Hot Springs. Cascade Springs is one of these thermal springs; the temperature at Cascade Springs averages 66°F. In fact, the temperature is so high that trout are unable to spawn in the water of Cascade Creek. Although the source of the heat producing these hot springs has not been identified with certainty, it appears to be related in one way or another to abnormally high temperatures beneath the surface in this area. These kinds of abnormal temperatures can be produced, for example, by conduction of heat toward the surface from cooling igneous rocks.

However, the presence of such an igneous rock mass beneath the Hot Springs area has not been confirmed.

An interesting side effect of the hot spring activity in the area of Cascade Springs, Hot Springs, and nearby Buffalo Gap, can also be observed. On both sides of the Fall River in downtown Hot Springs, one can see beds of conglomerate capping the upper surface of the valley. Examination of these rocks indicates that the conglomerate has been cemented into an extremely durable and resistant rock by precipitation of calcium carbonate around the boulders, pebbles, and sands. In all probability this calcium carbonate is a product of deposition from the hot springs issuing to the surface in this area. The distribution of these well-cemented conglomerates, which are probably Pleistocene in age, corresponds rather closely to the distribution of hot spring activities in the Black Hills.

GLOSSARY

Abrasion platform. A level, almost flat, submerged rock surface created by wave erosion.

Acidic. Of or relating to a silica-rich igneous rock composed primarily of light-colored minerals.

Aegirine. An uncommon, igneous rock-forming mineral of the pyroxene family NaFe $(SiO_3)_2$. Dark brown, green or black in color.

Agglomerate. A rock consisting of large (greater than 32 mm in diameter) irregular-shaped volcanic fragments in a matrix of smaller tuff and ash particles.

Albite. A common sodium-rich, gray or white mineral of the plagioclase-feldspar group $NaAlSi_3O_8$.

Alluvial. Of or relating to alluvium.

Alluvium. Loose sediment deposited by streams.

Amblygonite. An uncommon, white or green pegmatitic mineral rich in lithium, flourine, and phosphorus.

Amphibole. A group of common, dark-colored, iron and magnesium-rich silicate minerals distinguished by cleavage in two planes intersecting at angles of 56° and 124°.

Amphibolite. A metamorphic rock composed largely of amphibole. Sometimes the grains are aligned predominantly in one direction (lineation).

Andesite. A dark, fine-grained (aphanitic) igneous rock with a composition intermediate between rhyolite and basalt.

Angular unconformity. An unconformity between two rock units whose beds are not parallel.

Antecedent stream. An old stream which maintained its original course during and after uplift creating present topography.

Anticlinal core. The older rock at the center of an anticline.

Anticlinal nose. The place where the axis of a plunging anticline meets the land surface. The map pattern in the beds has arch-shaped appearance. The nose is the center of the arch.

Anticline. A convex-upward fold shaped like an arch.

Aphanitic. Of or relating to an igneous rock whose crystals are so small in size that they cannot be readily distinguished by the naked eye.

Apron (alluvial). A broad, gently sloping land surface composed of alluvium at the base of a mountain range.

Arch. See sea arch.

Arkose. A feldspar-rich sandstone.

Arroyo. A steep-walled, flat-bottomed channel of an intermittent stream in arid regions.

Ash. See volcanic ash.

Asphalt. A dark-brown to black, thick liquid formed from petroleum.

Asymmetrical fold. A fold having beds more steeply sloped on one side than on the other.

Atmospheric inversion. An increase in air temperature with elevation.

Attitude. The orientation of a bedding plane, usually expressed in terms of dip and strike.

Axis (fold). A line connecting the point in each bed which is most bent in the center of a fold.

Backwash. The seaward flow of water on a beach following the advance of a wave.

Badlands. Extremely rough topography characterized by steep slopes and sharp divides formed in arid regions.

Bank (submarine). A flat-topped mound on the sea floor.

Barranco. An especially deep, steep-walled canyon or arroyo. A cliff in an arid region.

Basalt. A fine-grained (aphanitic), dark igneous rock especially rich in iron and magnesium.

Base level. The land surface below which a stream cannot erode. Ultimate base level is sea level.

Basement. The crystalline rock onto which the sedimentary rock beds on the earth's surface were deposited.

Basic. Of or relating to an igneous rock low in silica and rich in iron and magnesium and composed largely of dark-colored minerals.

Batholith. A large (greater than 100 sq. km) pluton or group of plutons composed predominantly of large-grained (phaneritic), acidic igneous rocks.

Bay. A relatively large, open indentation or inlet into land by waters of the sea or a lake.

Baymouth bar (or barrier). A long, narrow bank of sand or gravel across the entrance to a bay.

Beach. The area composed of loose sand or pebbles between high- and low-tide levels along the shore of a standing body of water.

Beach ridge. A ridge of sand or pebbles built high on a shoreline beyond the present limit of tides or storm waves.

Bed. A layer of sedimentary rock having relatively uniform composition and bounded by bedding planes.

Bedding. The pattern and arrangement of beds of sedimentary rock.

Bedding plane. A planar surface marked by a visible break or change in rock type and separating beds of sedimentary rock.

Bedrock. The solid rock mass underlying soil or loose surface sediments.

Bench. A horizontal or gently sloping platform interrupting an otherwise steep slope.

Beryl. A fairly common, green or blue-green silicate mineral forming hexagonal prisms and often found in pegmatites.

Biotite. A very common, dark, iron and magnesium-rich mica mineral characterized by the ability to cleave or split readily into thin sheets.

Blowhole (marine). A near-vertical hole or fissure in the roof of a sea cave through which water is sprayed by waves at high tide.

Bluff. A high, steep-walled bank or cliff overlooking a broad, flat land area or body of water.

Brea. A liquid asphalt formed around an oil seep by evaporation of gas and vapor.

Breakaway-scarp. The steep cliff formed above a landslide, slump, or earthflow when the moving earth mass pulled away from the underlying, more solid surface material.

Breaker. A collapsing wave formed when waves approach a shoreline.

Breakwater. A man-made structure built to protect a harbor or beach from the force of waves.

Breccia. A volcanic or sedimentary rock containing large (greater than 2 mm in diameter), angular rock or mineral fragments.

Brink. The top of a steep slope, but not necessarily the highest part of the terrain.

Calcareous. Containing calcium carbonate ($CaCO_3$).

Calcite. A common, white or colorless mineral composed of calcium carbonate ($CaCO_3$). It has perfect rhombohedral cleavage.

Cañada. A ravine or narrow valley.

Canyon. A long, deep, relatively narrow, steep-sided valley.

Capture. See stream capture.

Carbonate rocks. Sedimentary rocks composed of the minerals calcite and dolomite.

Cassiterite. An uncommon, brown or black mineral used as the principal ore of tin.

Ceanothus. A common bush ("native lilac") of the chaparral community in California.

Chaparral. A type of native vegetation consisting of a dense growth of various scrubby plants on hillsides in areas with limited (12-25 inches) precipitation.

Chemical weathering. The chemical alteration of crystalline rocks on the earth's surface, usually in reaction to rainwater or ground water.

Chlorite. A common, green, platy silicate mineral containing iron and magnesium and usually found in metamorphic rocks.

Clastic. Of or relating to a sedimentary rock composed of rock fragments moved some distance from their place of origin.

Clastic dike. A sheet-like body of rock which cuts across a pre-existing rock mass. It is composed of fragments derived from underlying or overlying rock material.

Clay. (a) A rock fragment less than 1/256 mm in diameter. (b) A group of minerals having a strongly layered internal structure and formed by chemical weathering.

Claystone. A massive, homogeneous sedimentary rock composed largely of clay.

Cleavage. The splitting or breaking of a mineral along smooth planes defined by its internal crystal structure.

Clevelandite. A variety of albite forming fan-shaped arrangements of long, flat crystals and found in pegmatites.

Climbing dune. A dune formed as sand is driven upslope by the wind.

Clinker rock. Burnt-looking, jagged rock composed of a lava thrown out of a volcano or coal ash.

Closed depression. A basin or area of decreased elevation whose lowest point is completely surrounded by areas of higher elevation.

Coast (coastal zone). A land strip of indefinite width extending inland from the seashore to the first major topographic change.

Coastline. The line defined by the landward edge of the beach.

Colluvium. Fallen, loose rock debris which has accumulated at the base of a steep slope.

Columnar jointing. The breakage or fracture of igneous rock into polygonal columns during cooling.

Competent. A strong bed which does not break apart internally or change thickness when folded.

Concretion. A hard, compact, rounded mass formed in a void within sedimentary rock by chemical precipitation from groundwater.

Concretionary. Containing concretions.

Conglomerate. A clastic sedimentary rock composed of numerous large (greater than 2 mm in diameter), rounded fragments in a sand and silt matrix.

Contact. A surface between two different rock types.

Continental. Of or related to one of the earth's major land masses.

Core-stone. A roughly speroidal or ellipsoidal block of granite formed by subsurface weathering.

Correlation. The determination of stratigraphic positions of equal age (time planes) within sedimentary rocks occurring in separate locations.

Country rock. The prevailing rock surrounding and intruded by an igneous pluton.

Cove. A small indentation or bay within a coastline.

Cover head. Stream-laid rock debris which has accumulated on a stable marine terrace.

Creep. The slow, continuous downhill movement of soil and rock debris under the influence of gravity.

Crop out. The act of forming a bedrock exposure on the land surface.

Cross-beds. Especially thin sedimentary layers oriented at an angle to major bedding planes and usually contained within a single bed or sedimentary unit.

Cross-bedding. The pattern or arrangement of layers within a sedimentary rock having cross-beds.

Crust. The outermost solid layer of the earth.

Crustal block. One of the six coherent units into which the earth's crust is subdivided.

Crystal. A uniform, solid body whose atoms are arranged in a regularly repeating manner so that it is bound on all sides by smooth, planar surfaces.

Crystalline. Having the internal structure of crystals.

Crystalline rocks. Igneous and/or metamorphic rocks.

Cuesta. A hill with a gentle slope on one side and a steep slope on the other.

Cusp (beach). A low, crescent-shaped mound of sand on a beach formed by waves.

Debris. See rock debris.

Debris Avalanche. The sudden, rapid sliding of a mass of soil and broken rock down a steep slope.

Debris cone. A small, steep cone-shaped accumulation of rock fragments at the entrance of a gully or small canyon.

Debris flow. The rapid flow of a wet mass of mud and rock fragments downhill.

Deciduous. Plants (trees) that shed their leaves annually.

Decomposition. The chemical breakdown of rocks and minerals.

Dendritic drainage pattern. An arrangement of the courses of a river and its tributaries in which the streams diverge upstream irregularly in all directions, in a manner resembling the branches of a maple or oak tree.

Detritus. See rock debris.

Diatomite. A sedimentary rock composed almost exclusively of silica-rich skeletons of a single-celled algae.

Dike. A sheet-like igneous intrusion that cuts across the beds of the surrounding rock (country rock).

Diorite. A large-grained (phaneritic) igneous rock intermediate in composition between granite and gabbro.

Dip. The angle between a bedding plane (or fault) and a horizontal plane.

Dip slope. A smooth, inclined land surface conforming in direction and dip with the underlying rock beds.

Disconformity. An unconformity between two sedimentary rock units having parallel bedding planes.

Disintegration. The physical or mechanical breakup of rocks and minerals.

Distributaries. Channels which flow away from the main stream near the mouth or delta of a river.

Dolerite. A basalt composed of lath-shaped plagioclase feldspar crystals which are completely contained within pyroxene crystals.

Dolostone. A carbonate sedimentary rock composed of a mineral dolomite.

Dolomite. (1) A common, light-colored sedimentary rock-forming mineral ($CaMg(CO_3)_2$). (2) A carbonate sedimentary rock composed of the mineral dolomite (synonymous with dolostone).

Dome (structural). An anticline in which beds bend down in all directions.

Dome (topographic). A roughly circular, upwardly convex landform.

Drowned stream. A stream whose channel has been submerged by the sea near its mouth.

Dune. A low mound or bank of loose wind-blown sand.

Earthflow. Slow downslope movement of soil and rock debris lubricated with water.

Embayment. The formation of a bay.

Epicenter. The point on the earth's surface directly above subsurface point (Focus) where an earthquake shock originates.

Erosion. The mechanical loosening and chemical dissolution of rock materials from the earth's surface and their subsequent transportation.

Escarpment. A long cliff or steep slope separating two level or gently sloping areas usually formed by faulting.

Estuary. The seaward end of a river or an inlet of the sea which is influenced by tidal currents and by the intermixing of fresh and salt water.

Extrusion. The igneous process involving ejection of molten rock (magma) onto the earth's surface.

Extrusive rock. An igneous rock formed by extrusion.

Fan (alluvial). A cone-shaped deposit of rock debris formed by a stream when its slope changes suddenly as it leaves the mountains and enters more lowland areas.

Fanglomerate. An especially coarse, poorly sorted variety of conglomerate formed from consolidated **alluvial fan deposits.**

Fault. A break or fracture in the earth's crust along which blocks of rock on either side of the fracture have moved.

Fault plane. The planar surface along which a fault has moved.

Fault ridge. An elevated, elongate block of earth's crust lying between two roughly parallel faults.

Fault scarp. A long, straight cliff formed by movement along a fault.

Fault slice. A narrow block of rock wedged between two roughly parallel, closely adjacent faults.

Fault zone. A zone in the earth's crust consisting of many small, roughly parallel, closely spaced faults and fractures.

Feldspar. An abundant, light-colored group of rock-forming minerals composed of aluminum, silicon, oxygen, and one or more of the alkalies (sodium, calcium, and potassium).

Ferro-magnesian mineral. A mineral rich in iron and magnesium.

Flatiron. A triangular-shaped landform having a steeply inclined slope and formed by streams at the flank of a mountain.

Flood (stream). An especially high stream flow in which water flows over the banks confining the normal stream channel.

Flood plain. A strip of relatively smooth land bordering a stream subjected to episodic flooding.

Fluorite. An uncommon transparent mineral (CaF_2) with cube-shaped crystals which occurs in veins associated with lead, tin, and zinc ores.

Flute. A small, scoop-shaped groove in a rock surface created by water turbulence.

Flute cast. The silt or sand filling of a flute, best seen on the underside of an exposed sedimentary bed.

Fluvial. Of or related to streams.

Fold. A bend in a sedimentary bed.

Folding. The process of bending a sedimentary bed.

Foliation. The planar alignment or banding of minerals in a metamorphic rock.

Foot wall. The lower side of an inclined fault.

Foredune. The linear sand-dune ridge formed at the landward edge of a beach.

Foreland. An extensive area of land jutting out from the coastline into the sea.

Foreridge. A subsidiary ridge in front of a larger mountain mass.

Foreslope (beach). The steeply sloping portion of a beach subjected to the continuous advance and retreat of waves.

Formation. A readily distinguishable rock unit which can be mapped over a considerable area.

Fossil. The remains or traces of living matter which have been preserved by natural processes.

Fossil landscape. A terrain in which the principal landforms are relict from an earlier landscape.

Friable. Easily crumbled or crushable, as in a soft or poorly cemented sandstone.

Gabbro. A dark-colored, large-grained iron and magnesium-rich igneous rock composed primarily of pyroxene and plagioclase feldspar.

Garnet. A common, transparent red or brown silicate mineral lacking cleavage, rich in iron, magnesium and aluminum and formed in metamorphic rocks.

Geomorphology. The study of processes operating on the earth's surface. It deals especially with the description of landforms and the interpretation of their origin.

Geophysical exploration (methods). Subsurface exploration of rocks using indirect physical information such as gravity or magnetic variations to interpret rock composition and structure.

Geophysics. The study of the earth as a planet using principles derived from physics.

Glacier. A land-bound mass of ice and snow that persists throughout the year and flows downhill in response to gravity.

Glaucophane schist. A foliated metamorphic rock (schist) containing the blue mineral glaucophane (a sodium-bearing amphibole).

Gneiss. A large-grained metamorphic rock with irregular banding (foliation).

Gorge. An especially narrow, steep-walled stream valley, smaller in size than a canyon.

Gouge. Finely ground rock material surrounding a fault.

Graben. An elongate block of earth's crust which has dropped down between two steeply inclined gravity faults.

Granite. A common, large-grained, silica-rich igneous rock composed predominantly of quartz and feldspar.

Granitic. Of or related to an igneous rock composed of interlocking, equal sized feldspar and quartz crystals.

Granoblastic. Of or related to a metamorphic rock lacking foliation and composed of equal sized crystals.

Granodiorite. A large-grained igneous rock intermediate in composition between granite and diorite.

Graphite. An uncommon soft, gray or metallic appearing mineral, composed entirely of carbon and formed in veins surrounding igneous plutons.

Gravel. Loose sediments composed of particles larger than 2 mm. in size.

Gravity (or normal) fault. An inclined fault along which the block above the fault has moved down relative to the block below.

Gravity flow. The relatively rapid mass movement of rock debris downhill in response to gravity.

Groundmass. The fine material surrounding the large crystals (phenocrysts) in a porphyritic igneous rock.

Groundwater. Water which occupies all the void space in the rock beneath the water table. This underground water flows slowly in a downhill direction.

Grus. An accumulation of fragments derived from the mechanical and chemical breakdown of granite.

Gulch. A short, narrow, steep-sided ravine larger in size than a gully.

Gully. A small ravine.

Gypsum. A common, colorless or white, soft mineral composed of calcium sulphate and formed in sedimentary rocks in arid regions.

Hanging wall. The upper side of an inclined fault.

Headland. A projection of land into the sea.

High-tide Bench. A nearly flat platform at the base of a sea cliff, formed by wave movement during high tides.

Hogback. A steep ridge with a sharp summit formed from resistant, highly tilted rock beds.

Hoodoo. A bizarre-shaped column of rock or pinnacle created by erosion.

Hydrogen sulfide. A gas emitted from a volcano during an eruption or from a hot spring. It smells like rotten eggs.

Ice ages. The most recent time interval of earth history (i.e. the Pleistocene Epoch of the Cenozoic Era) during which especially large ice masses (glaciers) formed and moved over the continents.

Igneous. Of or related to igneous rocks.

Igneous rock. A rock formed by crystallization from a molten state (magma).

Immature sand. Sand consisting of fragments of various minerals in addition to quartz.

Impervious. Of or related to a rock through which fluids such as groundwater cannot flow.

Inclusion. A fragment of older country rock enclosed in an igneous rock.

Incompetent. Of or related to a weak rock bed which breaks or flows in response to pressure.

Inlet. A narrow water passage through a bar connecting a bay or lagoon with a larger body of water.

Intercalation. A layer of material laid down or inserted between two layers of different material.

Interference pattern. A pattern of intersecting wave crests on a water surface, created as waves approach an uneven shoreline.

Intermittent stream. A stream that flows only after a rainstorm or during the wet season.

Intraformational conglomerate. A conglomerate whose fragments are derived from a nearby sedimentary bed or rock unit.

Intrusion. An igneous rock body formed by the solidification of molten material (magma) which has been forced upward into pre-existing country rock.

Intrusive rocks. Igneous rocks formed in intrusions.

Inversion. See atmospheric inversion.

Iron Formation. A thin-bedded sedimentary rock unit usually of Precambrian age composed of at least 15% iron and formed by chemical precipitation of iron-rich minerals from seawater.

Joint. An almost vertical fracture in a rock along which no movement has occurred.

Jointing. A set or parallel joints cutting through a fairly uniform rock mass.

Kelp. Seaweed.

Kitchen midden. The garbage dump of an ancient Indian settlement.

Laccolith. An igneous intrusion (pluton) oriented roughly parallel to the beds of the surrounding country rock and having a flat lower surface and a rounded upper surface.

Lagoon. A shallow nearshore body of water separated from the ocean by a narrow beach or reef.

Landform. A topographic feature of the land surface which has a distinctive size and shape.

Land-laid. Sedimentary material deposited on land.

Landscape. Any association of landforms which can be seen in a single view.

Landslide. The downslope movement of a mass of rock and rock debris caused by slippage beneath the ground.

Lateral fault. A fault which has moved horizontally rather than vertically.

Latite. A porphyritic igneous rock composed primarily of plagioclase and orthoclase feldspar.

Lava. Molten rock material (magma) which has been extruded onto the earth's surface.

Left-lateral fault. A lateral fault along which the opposing block has moved to the left.

Lepidolite. An uncommon purple-colored mica mineral distinguished by its ability to break apart into sheets. It is found in pegmatites.

Leucocratic. Light-colored.

Lignite. A brown-black soft coal in which the plant material has been altered more than in peat but less than in bituminous coal.

Limb. One of the two sides of an anticline or syncline.

Limestone. A sedimentary rock composed largely of the mineral calcite and distinguished by its bubbling (effervescence) in response to hydrochloric acid.

Lineation. Any linear structure in a rock.

Lithology. The composition, size, shape, and arrangement of mineral grains in a rock.

Live oak (tree). An oak which retains green leaves throughout the year.

Longshore. Of or related to a process that moves water and debris laterally along a shoreline.

Longshore current (marine). A nearshore current that moves parallel to the shoreline.

Longshore drift (marine). The movement of sediment along a beach by a longshore current.

Magma. Molten rock material formed deep below the earth's surface. When cooled it forms igneous rocks.

Magnetite. An uncommon, metallic, iron-rich mineral distinguished by its strong magnetization.

Mantle rock. The layer of broken-up and altered rock debris overlying bedrock on the earth's surface.

Marble. A calcium-rich metamorphic rock which lacks foliation and is formed by recrystallizing limestone or dolomite.

Marine. Of or related to the ocean.

Mass movement. The downslope movement of a mass of rock or rock debris in response to gravity.

Matrix. The fine-grained particles in a rock which surround the larger particles.

Melange. A complex mixture of rock types produced by severe deformation.

Member. A subdivision of a formation having a distinct rock type (lithology) but occurring over a small area.

Mesa. A flat-topped hill or plateau bounded by steep cliffs.

Mesozoic. One of the eras of geologic time extending from 70 to 230 million years ago.

Metagraywacke. A dark metamorphic rock lacking foliation and formed by the recrystallization of a dark, large-grained sandstone, rich in feldspar, rock fragments, and having a mud-sized matrix.

Metamorphic rocks. Rocks that have been altered by metamorphism.

Metamorphism. The mineral and structural alteration of solid rock when subjected to heat, pressure, and chemically active fluids deep in the earth's crust.

Metavolcanic rocks. Usually dark metamorphic rocks lacking foliation and formed by the recrystallization of volcanic rocks.

Mica. A group of silicate minerals distinguished by their ability to readily split into thin, elastic sheets. This group includes biotite, lepidolite, and muscovite.

Microcline. A common form of the mineral orthoclase feldspar which is often found in granite and in pegmatites.

Mineral. A naturally occurring, inorganic substance having a definite chemical composition and crystal form.

Mineralized. Of or related to a rock having minerals that were introduced after rock formation, as in the emplacement of ores in veins.

Mio-Pliocene. An interval of geologic time extending from the uppermost Miocene Epoch into the lowermost Pliocene Epoch of the Cenozoic Era.

Mistletoe. A parasitic shrub that grows on some deciduous trees.

Monocline. A fold consisting of a sharp step-like bend in otherwise horizontal or gently dipping beds.

Monzonite. A fine-grained igneous rock composed largely of plagioclase and orthoclase feldspar.

Mudflow. The rapid downslope movement of a mass of mud lubricated with water.

Mudstone. A fine-grained sedimentary rock lacking bedding and composed of mixed clay, silt, and sand-sized particles.

Muscovite. A common, light-colored mica mineral distinguished by its ability to break apart readily into thin sheets.

Mussel. A dark-colored marine invertebrate animal enclosed by a shell having two valves. Mussels grow prolifically on rocks and pilings at water level near the shore.

Nepheline. An uncommon sodium and potassium-rich silicate mineral which is an important constituent of igneous rocks lacking quartz.

Nip (marine). A small notch or hollow at the base of a sea cliff caused by wave erosion at high tide.

Nodule (nodular). A small, hard mass or lump of mineral (or mineral aggregate) contained within a sedimentary rock bed.

Nonconformity. An unconformity between sedimentary rocks and older crystalline rocks (plutonic igneous or metamorphic rocks).

Normal fault. See gravity fault.

Nose. See anticlinal nose.

Oblique air photo. A photograph taken from an airplane with the axis of the camera tilted from the vertical.

Obsidian. A dark-colored, glassy volcanic igneous rock distinguished by its circular (conchoidal) fracture pattern.

Odometer. An instrument used for measuring distance.

Ore. A rock substance from which a mineral with economic value can be extracted.

Orogeny. The process of mountain formation involving folding, faulting, and thrusting.

Orthoclase. A common, light-colored potassium-rich silicate mineral of the feldspar group which is an important constituent of granite or pegmatite.

Outcrop. An exposure of bedrock on the surface.

Outlier. A small rock mass that is detached from another larger mass of the same rock.

Overturned beds. Sedimentary beds which have been folded so that older beds lie on top of younger beds.

Overturned fold. A fold in which the beds of one limb are turned upside down.

Paleozoic. One of the eras of geologic time extending from 230 to 600 million years ago.

Parting. A surface along which a rock readily splits into layers.

Pediment. A broad, gently sloping plain created by erosion at the base of a mountain range in an arid region.

Pegmatite (pegmatitic). An especially large-grained igneous rock having a composition similar to granite. Pegmatites are commonly formed at the margins of batholiths.

Percolation. The flow of water downward through small holes in rock.

Perthite. An uncommon form of feldspar consisting of an intergrowth of the minerals, microcline and albite.

Phaneritic. Of or relating to an igneous rock whose crystals are so large in size that they can be readily distinguished by the naked eye.

Phenocryst. A large, conspicuous crystal set in a fine-grained groundmass in a porphyritic igneous rock.

Pholad boring. A hemispherical or tubular hole in a seashore stone, formed by the boring of certain clam-like invertebrates.

Phonolite. A gray, fine-grained volcanic igneous rock composed primarily of the minerals, orthoclase feldspar and nepheline.

Phosphate rock. A sedimentary rock composed largely of phosphate minerals (i.e. minerals formed from a compound of phosphorous and oxygen).

Phyllite. A fine-grained, foliated metamorphic rock containing mica and chlorite and distinguished by its silky sheen.

Physical weathering. The mechanical breakdown of crystalline rocks on the earth's surface, usually in response to frost action and temperature changes.

Piedmont. Of or relating to the land area at the foot of a mountain range.

Piracy. See stream capture.

Placer. A mineral deposit formed by the accumulation of eroded debris (for example, a concentration of gold fragments in stream sands and gravels).

Plagioclase. A common, white or gray, calcium and/or sodium-rich silicate mineral of the feldspar group which is an important constituent of basic igneous rocks such as basalt.

Plain. A broad land area at a low elevation having little or no topographic relief.

Plate tectonics. The movement and interaction of large blocks of crust (plates) on the earth's surface.

Playa. A flat, smooth dry or intermittent lake in arid regions.

Plio-Pleistocene. An interval of geologic time extending from the uppermost Pliocene Epoch into the lowermost Pleistocene Epoch of the Cenozoic Era.

Plug. A vertical, cylindrical igneous body representing the subsurface passageway for magma up into a former volcano.

Plunge (Plunging). The angle between the axis of a fold and the horizontal. A fold is said to be plunging if this angle is not equal to 0°.

Pluton (plutonic). An igneous rock body which has been forced up as magma into surrounding subsurface country rock (synonymous with intrusion).

Pore. A small void space within a rock.

Porphyry (porphyritic). An igneous rock containing large conspicuous crystals in a fine-grained matrix.

Potassium-argon. A method of age dating rocks using a ratio between the amount of the radioactive elements, potassium and argon, contained within a rock.

Precambrian. The oldest era of geologic time extending from the earth's formation approximately 5 billion years ago to 600 million years ago.

Precipitation. (a) The discharge of water from the atmosphere onto the earth's surface. (b) The process whereby solid materials form as a result of chemical reactions in a fluid.

Pumice. A porous volcanic igneous rock composed of irregularly arranged glass shards.

Pyroclastic. Of or relating to ash and debris thrown out of a volcano.

Pyroxene. A group of common, dark-colored, iron and magnesium-rich silicate minerals distinguished by cleavage in two planes intersecting at right angles.

Quartz. An abundant, light-colored, transparent silicate mineral distinguished by its high hardness and its lack of cleavage.

Quartzite. A metamorphic rock lacking foliation and composed entirely of the mineral quartz.

Quartz latite porphyry. A light-colored porphyritic igneous rock having a groundmass intermediate in composition between rhyolite and dacite and containing plagioclase and quartz phenocrysts.

Quartz monzonite. A large-grained igneous rock composed largely of quartz and feldspar but intermediate in composition between granite and granodiorite.

Radioactive. Of or relating to the ability of some elements to change spontaneously into other elements by emitting charged particles.

Radiocarbon. The radioactive form of carbon which disintegrates at a known rate and is especially useful in age dating material younger than 70,000 year old.

Rainbeat. The impact of raindrops on the ground.

Reef. A submerged mound or ridge usually formed by the accumulation of plant and animal skeletons.

Refraction (wave, swell). The change in direction or bending of a wave front as its velocity changes.

Regression. See regressive shoreline.

Regressive shoreline. A shoreline that moves slowly seaward as a result of a fall in sea level or uplift of the land.

Relief. (a) The difference in elevation between the high and low points of a land surface. (b) The general character (its shape, unevenness) of a land surface.

Revolution. An episode of mountain building or of severe folding and faulting.

Rhyolite. A fine-grained, light-colored igneous rock composed largely of orthoclase feldspar and quartz.

Rib. A narrow, ridge-like outcrop of rock extending down a steep hillside.

Rift. A topographic trench or fissure (about 3 km wide) along a fault.

Right-lateral fault. A lateral fault along which the opposing block has moved to the right.

Rill. A narrow channel directed straight downhill which was formed by an intermittent stream.

Rilled. Of or relating to a slope having many small, parallel rills.

Ring dike. An igneous dike with a ring-like shape.

Riparian. Of or pertaining to a stream bank.

Rip current. A strong, narrow surface current flowing seaward from the shore.

Rock. An aggregate of minerals.

Rock debris (detritus). Surficial broken-up rock material.

Rockfall. The sudden fast downslope movement or free fall of a rock mass from a steep hillside.

Rock slide. The fast downslope movement of a rock mass by sliding along its base.

Rock-slide debris. The broken-up rock formed by a rock slide, rockfall, or rock avalanche.

Roundstone. A rock fragment rounded by abrasion and wear during movement by currents or waves.

Saddle. A low point on a mountain ridge crest or a broad pass across a ridge.

Sag pond. A pond in a depression along a fault.

Sand flow. The subaerial or submarine flow of a loose sand mass.

Sandstone. A clastic sedimentary rock composed of consolidated sand-sized (1/16–2 mm in diameter) particles.

Sandstone dike. A layer of sand that cuts across the bedding of the pre-existing rock and is derived from underlying or overlying material.

Scarp. A long, steep, straight cliff, often produced by faulting.

Schist. A medium or large-grained, strongly foliated metamorphic rock with abundant parallel-oriented mica flakes.

Schistosity. A type of foliation characteristic of schists in which mineral grains have a parallel, planar arrangement.

Scour channel. A small, shallow channel in a stream bed or on the ocean floor created by the strong digging action of currents.

Sea arch. A tunnel or opening through a headland created by wave erosion.

Sea cliff. A cliff formed by wave action.

Sea-floor spreading. The process whereby new seafloor is formed from molten subsurface rock at the boundary between two crustal blocks (plates) as they gradually move apart.

Sediments (sedimentary). Solid rock fragments that have been moved from their place of origin or that have formed from chemical reactions in a fluid (i.e. by precipitation).

Sedimentary rock. A rock formed from the consolidation of loose sediment.

Sedimentary structures. Geometric features in sedimentary rocks formed by layers of grains of different sizes, shapes, and compositions during sediment deposition (for example, ripple marks, bedding).

Seismology. The study of earthquakes and waves produced by earthquakes.

Septum. An older mass of metamorphic rock separating two nearby igneous plutons.

Serpentine or serpentinite. A dark-colored, fine-grained, unfoliated metamorphic rock composed of minerals of the serpentine group which are formed by the alteration of pyroxene and other minerals rich in iron and magnesium.

Shale. A clastic sedimentary rock composed of consolidated silt and clay-sized (less than 1/16 mm diameter) particles.

Sheetflow. An overland flow of water downhill in a thin continuous sheet.

Shore. A narrow strip of land bordering a water body over which water is moved by tides and waves.

Shoreline. A shifting line defined by the intersection of water and land along a shore.

Shoreline processes. Transportation, erosion, and deposition of sediments along a shore by waves, and tides.

Silica. The chemical compound of silicon and oxygen (SiO_2).

Silicate. Of or relating to a mineral composed of silicon and oxygen linked together into the unit SiO_4.

Silicic. Rich in silica.

Silicified. Impregnated with or replaced by silica.

Sill. A sheet-like igneous intrusion that parallels the beds of the surrounding rock (country rock).

Sillimanite. A green or white silicate mineral distinguished by the needle-like shape of its crystals. It is common in metamorphic rocks that have been subjected to very high temperatures.

Siltstone. A clastic sedimentary rock composed of consolidated silt-sized (1/16-1/256 mm diameter) particles.

Skin slide. The slippage of a thin layer of water-saturated soil on a steep, often grassy slope.

Slate. A fine-grained, strongly foliated metamorphic rock which readily breaks into numerous, smooth, parallel surfaces.

Slickenside. A polished and scratched surface created by slippage along a fault plane.

Slide. See landslide.

Slope wash. A mass of loose soil and rock debris carried downslope by water and not confined in a channel.

Slump. A landslide involving the downward slippage and rotary movement of a rock mass along a curved surface.

Soil. The layer of surficial rock debris that has been so altered by weathering that it can support rooted plants.

Soil slip. See skin slide.

Sorting. A measure of the size range of particles in a sedimentary rock.

Spanish grant. A large tract of land granted to an individual by the government of Spain within its American provinces.

Spodumene. An uncommon, lithium-rich, white, purple, or green mineral of the pyroxene family distinguished by its large prismatic crystals. It is often found in pegmatites.

Spit. A shoreline sandbar partly closing the entrance to a bay or estuary.

Spur. A small ridge which projects at right angles to the crest of a hill or mountain.

Stack. A small, steep-sided, pillar-like rocky island detached from a nearby cliff on the shore by wave erosion.

Stalactite. A conical calcareous deposit hanging down from the roof of a cave. It is precipitated from groundwater dripping from the roof.

Stalagmite. A conical calcareous deposit projecting upward from the floor of a cave. It is formed from water dripping down from a stalactite.

Staurolite. A brown or black silicate mineral forming twinned crystals which often have the shape of a cross. It is common in metamorphic rocks that have been subjected to moderate temperatures and pressures.

Stock. An igneous intrusion (pluton) which resembles a batholith but is less than 100 sq. km in surface exposure.

Storm-wave bench. A narrow, horizontal platform at the base of a sea cliff formed by wave erosion during storms.

Strand (marine). The narrow strip of land bordering the ocean (synonymous with beach).

Strata. Sedimentary rock beds.

Stratigraphic. Of or relating to the arrangement and succession of sedimentary beds.

Stratigraphy. The study of the form, arrangement, and distribution of rock beds.

Stream capture. The diversion of the headwaters of one stream by the headward growth of another stream.

Stream-gravels. The gravel transported and deposited by a stream.

Strike. The compass bearing or orientation of a horizontal line on an inclined plane such as a bedding plane or fault.

Stripping. The erosion of overlying material to expose the bedding plane of an underlying bed.

Structural trough. A low-lying, trough-like land feature formed by faulting or folding.

Structurally depressed. A region of lower elevation caused by faulting or folding.

Structure. The general orientation and arrangement of rock masses in an area with special emphasis on folds, faults, and fractures.

Submarine canyon. A large, steep-sided valley or canyon carved into the seafloor.

Superimposed stream. A stream that has cut down through the surface into rocks of different character and structure.

Superimposition. See superimposed stream.

Surf zone. The area along the shore across which breakers are active.

Swale. A small topographic depression in otherwise level land.

Swash. The landward flow of water on a beach caused by the advance of a wave.

Syncline. A U-shaped, concave-upward fold.

Taconite. An iron-rich amorphous silicic sedimentary rock. An Iron Formation which can be mined for iron ore when weathered.

Talus. An accumulation of loose, fallen rock fragments at the base of a cliff.

Tantalite-columbite. A rare, black niobium and tantalum-rich mineral sometimes found in pegmatites.

Tar. A dark-brown to black, thick liquid formed from petroleum. It flows more readily than asphalt.

Tectonic. Of or pertaining to large-scale deformation (folding and faulting) in the Earth's crust.

Telluride. A rare mineral compound formed from telluride and a metal such as silver.

Terrace. Any long, narrow, level or gently inclined surface interrupting an otherwise steep slope. A terrace is larger than a bench but smaller than a plain.

Terrace gravel. An accumulation of gravel on a terrace usually deposited by streams.

Terrain. A land area under observation at any particular time.

Terrane. A land area in which a particular rock formation prevails.

Terrestrial. (a) Pertaining to the planet Earth. (b) Pertaining to the earth's dry land.

Tertiary. An interval of geologic time within the Cenozoic Era extending from 70 to 2 million years ago.

Threadflow. The flow of water in small ill-defined, interlaced channels on the upper part of a slope.

Thrust fault. A gently inclined fault along which one block is thrust up over another.

Tonalite. A large-grained igneous rock composed primarily of quartz, plagioclase, feldspar, and amphibole (synonymous with quartz diorite).

Tourmaline. A fairly common silicate mineral often black in color with distinctive long hexagonal crystals. It is commonly found in pegmatites.

Tourmalinized. Of or relating to a rock with abundant tourmaline.

Trachyte porphyry. A gray porphyritic igneous rock composed mostly of plagioclase feldspar.

Transgressive shoreline. A shoreline which moves gradually landward because of the rising of sea level or the sinking of land.

Travertine. A hard, massive calcareous sedimentary rock precipitated from groundwater and commonly found in caves.

Trellis (drainage pattern). A rectangular arrangement of stream channels which have parallel and perpendicular courses.

Tuff. See volcanic tuff.

Tule. A swamp or marsh plant, more commonly known as cattail.

Turbidite. Sediment deposited by a turbidity current.

Turbidity current. An underwater bottom current caused when a mass of fine debris mixed with water moves suddenly and rapidly down a steep slope.

Turbidity flow. See turbidity current.

Turbulence. A type of water flow in which particles of water move in all different directions.

Two-story valley. A younger valley or canyon incised within remnants of an older, wider valley.

Unconformity. A surface of erosion or non-deposition separating two rock beds.

Undertow. The intermittent seaward flow of water along an ocean bottom away from a beach.

Vein. A sheet-like deposit of minerals along a fracture.

Vertical air photo. A photograph taken from an airplane with the axis of the camera pointed straight down toward the ground.

Volcanic ash. Loose, fine-grained (less than 3 mm diameter) igneous debris explosively erupted from a volcano.

Volcanic cinders. Loose, large-grained (3-25 mm diameter) porous igneous debris explosively erupted from a volcano.

Volcanic tuff. A compacted deposit of igneous debris consisting of ash, cinders, and larger fragments of solid volcanic rock.

Volcanism. The expulsion of molten igneous rock and pyroclastic debris onto the earth's surface through a volcano.

Volcano (volcanic). A circular opening or vent in the earth's surface through which molten igneous rock, pyroclastic debris, and gas erupt.

Vug. A small cavity in a rock filled with crystals of a different composition.

Warp. A broad bend in the earth's crust.

Water table. The level beneath the ground surface below which all void spaces in rocks are filled with water.

Wave (marine). An oscillatory movement of water as seen in the rise and fall of a water surface.

Wave (or swell) front. The linear line defined by the forward movement of a wave crest.

Wave height. The vertical distance between the crest and trough of a wave.

Wave length. The horizontal distance between two adjacent wave crests.

Wave ray. A line drawn perpendicular to a wave front.

Weathering. The mechanical breakup and chemical decomposition of rock materials on or near the earth's surface.

Weeping cliff. A steep rock face through which water continuously seeps in a thin film.

Whaleback. A large elongate mound or hill having a smooth rounded surface.

Wildfire. A natural fire in forest, chaparral, brush, or grass.

Wolframite. A brown or grayish-black mineral used as a principal ore of tungsten.

Wool-sack boulders. Large, rounded boulders formed on hillslopes above massive jointed rocks.

REFERENCES

Agenbroad, L. D. 1977. Mammoth Site of Hot Springs, South Dakota. *Hot Springs, Mammoth Site of Hot Springs South, Dakota, Inc.* 20 p.

Agenbroad, L. D. 1978. "Excavation of the Hot Springs Mammoth Site: field seasons 1974-77." *Friends of the Pleistocene Guidebook,* p. 1-11.

Cameron, E. N.; Jahns, R. H.; McNair, A. H.; and Page, L. R. 1949. Internal structure of granitic pegmatites. *Econ. Geology Mon. 2,* 115 p.

Darton, N. H. 1912. Volcanic action in the Black Hills of South Dakota. *Science n.s.,* v. 36, p. 602-603.

Darton, N. H., and Paige, S. 1925. Description of the central Black Hills (S. Dak.) *U.S. Geol. Survey Atlas, Folio 219.*

Gries, J. P. 1975. Paleozoic rocks. *U.S. Geol. Survey Rept. to 94th Congress,* p. 32-38.

Harrer, C. M., 1966. Iron resources of South Dakota. *U.S. Bur. Mines Inf. Circ. 8278,* 160 p.

Hess, F. L. 1925. The natural history of the pegmatites. *Eng. Mining Jour.* v. 120, p. 289-298.

Homestake Mining Company. 1976. Homestake Centennial. *Homestake Mining Company,* Lead, South Dakota, 96 p.

Jahns, R. H. 1953. The genesis of pegmatites. *Am. Mineralogist,* v. 38, p. 563-598, 1078-1112.

Kirchner, J. G. 1977. Evidence for late Tertiary volcanic activity in the northern Black Hills, South Dakota. *Science,* v. 196, p. 977.

Laury, R. 1978. Sedimentology of the Hot Springs Mammoth Site: a preliminary report. *Friends of the Pleistocene Guidebook,* p. 12-26.

Lisenbee, A. L. 1975. Black Hills. *U.S. Geol. Survey Rept. to 94th Congress,* p. 54-56.

Noble, J. A. 1950 Ore mineralization in the Homestake gold mine, Lead, South Dakota. *Geol. Soc. America Bull.,* v. 61, p. 221-252.

Noble, J. A. 1950. Ore mineralization in the the forcible intrusion of magma. *Jour. Geol.,* v. 60, p. 34-57.

Norton, J. J. 1975. Pegmatite minerals. *U.S. Geol. Survey Rept. to 94th Congress,* p. 132-149.

Norton, J. J., and others, 1964. Geology and mineral deposits of some pegmatites in the southern Black Hills, South Dakota. *U.S. Geol. Survey Prof. Paper 297-E,* p. 293-341.

Norton, J. J., and Redden, J. A., 1975. Gold and silver deposits. *U.S. Geol. Survey Rept. to 94th Congress,* p. 78-90.

Powell, J. E., Norton, J. J., and Adolphson, D. G. 1973. Water resources and geology of Mount Rushmore National Memorial, South Dakota. *U.S. Geol. Survey Water Supply Paper, 1865,* 50 p.

Rahn, P. H., and Gries, J. P. 1973. Large springs in the Black Hills, South Dakota and Wyoming. *S. Dak. Geol. Survey, Rept. Invest. 107,* 46 p.

Ratté, J. C. and Wayland, R. G. 1969. Geology of the Hill City quadrangle, Pennington County, South Dakota—a preliminary report. *U.S. Geol. Surv. Bull. 1271-B,* p. 1-14.

Ratté, J. C. and Zartman, R. E., 1970. Bear Mountain gneiss dome, Black Hills, South Dakota—age and structure. *Geol. Soc. America Abs. with Programs,* v. 2, p. 345.

Redden, J. A. 1963. Geology and pegmatites of the Fourmile quadrangle, Black Hills, South Dakota. *U.S. Geol. Survey Prof. Paper 297-D*, p. 199-291.

Redden, J. A. 1968. Geology of the Berne quadrangle, Black Hills, South Dakota. *U.S. Geol. Survey Prof. Paper 297-F*, p. 343-408.

Redden, J. A. 1975a. Iron. *U.S. Geol. Survey Rept. to 94th Congress*, p. 95-98.

Redden, J. A. 1975b. Precambrian geology of the Black Hills. *U.S. Geol. Survey Rept. to 94th Congress*, p. 21-28.

Redden, J. A. 1975c. Tertiary igneous rocks. *U.S. Geol. Survey Rept. to 94th Congress*, p. 45-47.

Shapiro, L. H. 1971. Structural geology of the Fanny Peak lineament, Black Hills, Wyoming-South Dakota. *Wyoming Geol. Assoc. Guidebook, 23d Ann. Field Conf.*, p. 61-64.

Sheridan, D. M., Stephens, H. G., Staatz, M. H., and Norton, J. J. 1957. Geology and beryl deposits of the Peerless pegmatite, Pennington County, South Dakota. *U.S. Geol. Survey Prof. Paper 297-A*, p. 1-47.

INDEX

Aegirine, 30, 83, 86, 94
Albite, 20, 39, 126, 127, 130, 157
Algae, 63, 77, 79
Algal mats, 63
Alladin Formation, 72-75
Amblygonite, 39, 120
Amphibolite, 24
Ankerite, 92
Arsenopyrite, 97
Attitudes, of rocks, 11-13

Badlands National Monument, 120
Badlands region, 10
Basalt lava flows, 23, 24
Bear Butte, 51-56
  igneous intrusion at, 11, 18, 30, 49, 51-56
  from Sly Hill, 56
  from Terry Peak, 85
Bear Butte Circus, 55-56
Bear Butte Creek, 59
Bear Butte road, 56
Bear Mountain, 20, 24
Bears, 161
Beaver Creek, 158
Belemnites, 28, 45
Belle Fourche River, 17, 18, 117
Beryl, 38, 39, 120, 139
Beryllium, 39
Bighorn Mountains, 1, 11, 75
Biotite, 23
  at Bear Mountain, 20
  of Bull Moose Pegmatite, 155
  of Custer schist, 149-150
  at Cutting Stock, 80
  at Etta Mine, 126
  in Lead area, 86, 92, 97
  at Mount Rushmore, 133, 136
  at Nemo, 20

at Peerless Mine, 122
at Strawberry Ridge, 104
Birds, 161
Bivalves, 28, 29, 45, 161
Black Hills Uplift, 11, 27, 29, 59
  and Bear Butte, 55
  and Crow Peak, 46
  and Dakota Hogback, 18
  from Dinosaur Overlook, 117
  and folding, 63, 65
  and Racetrack, 17
  from Sly Hill, 56
  from Terry Peak, 85
Bland, James, 155
Borglum, Gutzon, 136
Boudinage, 105
Boulder Canyon, 59
Boulder Park, 59, 63
Boulder Park Country Club, 63-65
Boulder Park Syncline, 59, 63-65
Boxelder Creek, 109
Brachiopods, 27, 72, 112-115
Bridal Veil Falls, 46-51
Buffalo Gap, 112, 157-161, 169
Bull Moose Pegmatite, 153-157

Calcite, 17, 79, 148-149
Calcium carbonate, 17, 79-80
  at Buffalo Gap, 158, 161
  at Hot Springs, 169
  in Jewel Cave, 146, 148, 149
  in Little Elk Creek Canyon, 115
  on Sly Hill, 56
Cambrian age, 9, 24-27, 75
  at Bridal Veil Falls, 46
  at Harney Peak, 22
  in Lead, 67, 94
  at Little Elk Creek Canyon, 112

at Nemo, 109
at Strawberry Ridge, 100-101
under Terry Peak, 86
at Whitewood Creek, 67
Camels, 161
Camptonectes bellistriatus, 45
Carlile Formation, 54
Cascade Creek, 168
Cascade Springs, 158, 164-169
Case-hardening, 56
Cassiterite, 39, 120, 125
Caves, 17, 80, 144-149, 167
Cenozoic age, 10-11, 29, 30-34
  at Bear Butte, 11, 51
  at Bridal Veil Falls, 49
  of Cutting Stock, 80
  in Lead area, 88, 94, 96
  of Sheep Mountain Stock, 65
  in Strawberry Ridge, 104
  under Terry Peak, 86
  at Tomahawk Lake Country Club, 105-109
Central City, 80
Central Crystalline area, 15
Cephalopods, 27, 28, 29, 77
Chadron, Nebraska, 161
Chadron State College, 161
Cheyenne Crossing, 83
Cheyenne River, 17, 18, 117
Chicago and Northwestern Railroad, 65
Chlorite, 86, 92, 97
City Creek, 65
Clams, 161
Cleavelandite, 127
Climate, 13
Columbite-tantalite, 120, 157
Confederate memorial, 136
Conodonts, 76
Coral, 27, 77, 79, 112-115
Corning Glass Company, 124
Coyotes, 161

Creep, 105
Cretaceous age, 7, 18, 28-30, 54, 59, 120
    of Fall River Formation, 10, 56
    of Lakota Formation, 10, 56
    of Sundance Formation, 28
Crinoids, 45
Crook Mountain, 11
Crow Peak, 46
Crown Mine, 37-38
Cummingtonite, 97
Custer County, South Dakota, 157
Custer, George A., 85
Custer, South Dakota, 7, 34, 37-38, 40, 144, 149-153
Cut-and-fill method, 99
Cutting Mine, 82
Cutting Stock, 65, 80-83, 96

Dakota Hogback, 18, 56, 85, 117
Dakota Quartz Products, 124, 125
Darrow, Donald, 125
Deadwood Creek, 80
Deadwood Formation, 15, 24-27, 70-75
    at Bridal Veil Falls, 46, 49
    in Lead, 94, 96
    of Little Elk Creek Canyon, 112-115
    at Nemo, 109
    at Strawberry Ridge, 100-104
    at White Gate, 112
    at Whitewood Creek, 67
Deadwood Gulch, 40
Deadwood, South Dakota, 30, 31, 33, 65, 70, 75
    Cenozoic volcanic rocks near, 105
    Sheep Mountain Stock near, 65
    Strawberry Ridge near, 100, 104
    Whitewood Creek near, 65
Devonian age, 9, 27, 77
Dinosaur Overlook, 15, 117-120
Dinosaurs, 10, 28, 29
Dogtooth spar, 149
Dolerite, 23
Dolostone, 9, 24, 70, 76, 94
Doming, 11-15, 18, 24, 30-31, 85, 129-130
Dripstone, 149
Dunes, 115
Dunlop Avenue, Deadwood, 70

Echinoderms, 79
Elkhorn Peak, 18, 49, 85
Ellison Formation, 86-88, 94, 96
Englewood Formation, 27, 75-76, 77-79, 112, 115
Eocene Epoch, 33, 49, 51, 109
Erosion, 49-51, 86, 116-117, 144. *See also* Flooding
Etta Mine, 34, 38, 124-129
Etta Pegmatite, 34, 38, 124-129
Evanston Avenue, Hot Springs, 161

Fall River, 161, 169
Fall River Formation, 10, 17, 18, 55, 56
Faults, 94
Feldspars, 20, 23, 30, 34, 38-39
    at Bear Mountain, 20
    of Cutting Stock, 82
    at Etta Mine, 34, 125, 126
    at Hugo Mine, 38
    of Iron Mountain Pegmatite, 139
    at Mount Rushmore, 130, 133
    near Nemo, 20
    at Peerless Mine, 120, 122
    of Sheep Mountain Stock, 65
    of Terry Peak, 86
Felsite, 88
Fish, 27, 59
Flooding, 28, 49-51, 75, 116, 117-120
Flowstone, 149
Fluorite, 82
Folding, 63, 65, 104-105, 133
Fort Union Group, 29
Fossils
    of Alladin Formation, 75
    in Deadwood Formation, 27, 46, 72, 75, 112-115
    in Englewood Formation, 27, 77, 79
    of Graneros Formation, 59
    in Hell Creek Formation, 29
    at Hot Springs, 161-164
    of Laramide Orogeny, 29
    in Minnekahta Formation, 28, 63, 161
    in Minnelusa Formation, 28
    of Morrison Formation, 10, 28
    in Opeche Formation, 28
    in Pahasapa Formation, 27, 79, 112-115, 146
    of Sundance Formation, 10, 28, 45

    of Unkpapa Sandstone, 10
    of Whitewood Formation, 76-77
    of Winnipeg Formation, 76
Fox Hills Sandstone, 10
French Creek, 155
Fuson Member, 10, 56

Gabbro, 23
Galena, 30
Garnets, 23
    of Custer schist, 149-150, 153
    near Lead, 86, 97, 104
    in Mount Rushmore, 132
    of Peerless Pegmatite, 122
Gastropods, 29, 77
Glauconite, 46, 72-75
Glory-hole mining method, 129
Gneissosity, 20
Gold, 39-40
    of Custer schist, 153
    at Cutting Mine, 82
    Homestake, 39-40, 85, 86, 94-100
    at Peerless Mine, 124
    at Whitewood Creek, 65-70
Gold telluride minerals, 82
Graneros Formation, 59
Granite, 20-22, 34. *See also* Harney Peak Granite; Pegmatites, granitic
Graphite, 92
Great Plains, 1, 17, 59, 85, 120
Gypsum, 9, 28, 43, 149, 167
Gypsum Spring Formation, 9-10, 28, 43

Halysites, 77
Harney Peak, 15, 20, 83, 85, 117
Harney Peak Granite, 7, 20-22, 23, 24, 31, 34
    at Harney Peak, 20, 85, 117
    on Iron Mountain, 139
    on Mount Baldy, 139
    and Mount Rushmore, 20, 129, 130-133, 136
    in The Needles, 20, 22, 139-144
Hell Creek Formation, 10, 29
Hematite, 101, 104, 109
Highway 14, U.S., 49, 59-61, 65, 75
Highway 14A, U.S., 46, 49, 65, 70, 80, 83, 92
Highway 16, U.S., 144, 149
Highway 16A, U.S., 120, 125, 129, 139, 153

Highway 18, U.S., 157, 164
Highway 71, South Dakota, 164
Highway 79, South Dakota, 51, 157
Highway 85, U.S., 65, 70, 75, 83, 86, 88, 94
Highway 87, U.S., 139
Highway 89, 149, 153
Highway I-90, U.S., 43, 112
Highway 101, Custer County, 157
Highway 244, 129, 132
Highway 385, U.S., 100, 104, 105, 161
Hill City, 7
Hoffman, Morris, 65
Homestake Formation, 86, 92, 94-100
Homestake Mine, 30, 39-40, 67, 69, 85, 97-100
Homestake Mining Company, 100
Hornblende, 24, 80-82, 133
Hot springs, 157, 158, 164-169
Hot Springs, South Dakota, 112, 157, 158, 161-164, 168, 169
Houston Street, Lead, 94
Hugo Mine, 38, 124

Intertonguing, 28
Intrusions, igneous, 11, 18-20, 23, 30-31, 33
    at Bear Butte, 11, 18, 30, 49, 51-56
    at Bridal Veil Falls, 46-49
    of Cutting Stock, 80-82
    in Lead area, 30, 31, 39, 88, 96
    of Sheep Mountain Stock, 65
    and Terry Peak, 85-86
Iron, 24, 41, 100-104, 105, 109-112, 133
Iron Mountain, 139

Jewel Cave National Monument, 144-149
Jointing, 15, 146
Junction Street, Sturgis, 56
Jurassic age, 10, 28, 157-158

Keith (R. H.) Park, 164
Keystone, South Dakota, 120, 122, 124-125, 129, 139

Laccoliths, 30, 86
Lake Superior, 109

Lakota Formation, 10
    of Bear Butte Circus, 55
    of Dakota Hogback, 18
    from Dinosaur Overlook, 117
    of Lookout Peak, 43, 45
    of Red Valley, 17, 46
    on Sly Hill, 56
Landsliding, 17, 49
Laramide Orogeny, 11, 13, 17, 29, 56, 146
Latite, 30
Lead Dome, 11, 31, 85
Lead, South Dakota, 7
    gold in, 39, 67, 94, 97
    igneous intrusions of, 30, 31, 39, 88, 96
    Precambrian rocks around, 86-94
    stocks of, 65, 96
    and Strawberry Ridge schist, 104
    Terry Peak near, 83, 85
Lepidolite, 39
Limestone, 9, 13
    of Englewood Formation, 77-79
    of Laramide Orogeny, 28-29
    of Minnekahta Formation, 9, 46, 61-63, 117, 167
    of Minnelusa Formation, 61
    of Minnewaste Member, 10
    of Pahasapa Formation, 17, 54, 61, 79, 85-86, 146
Limestone Plateau, 15-17, 85
Lithium, 38, 39, 125
Little Big Horn, 85
Little Elk Creek Canyon, 112-117
Little Elk Creek Monocline, 112, 116
Locality 8A, 70
Locality 8B, 70
Locality 12A, 86-88
Locality 12B, 88-92, 94
Locality 12C, 92-94
Locality 12D, 94
Locality 13A, 94, 96
Locality 13B, 94
Locality 16A, 106
Locality 16B, 106
Lookout Peak, 43-46

Maclurites, 77
Magnetite, 109
Main Street, Deadwood, 70
Main Street, Rapid City, 117
Mammoths, 161-164
Manganese dioxide, 149

Marble, dolomitic, 24
Martite, 109
Mercury, 67
Mesozoic age, 7, 30
    of Dakota Hogback, 117
    of Racetrack, 9, 17, 117
    of Spearfish Formation, 9, 28, 43-46, 117
Metamorphism, 20, 23, 24, 31, 136
    of Custer schist, 153
    in Lead area, 86, 97
    at Mount Rushmore, 133
    at Peerless Mine, 122
Micas, 24, 39
    at Bear Mountain, 20
    of Bull Moose Pegmatite, 155
    at Crown Mine, 38
    of Custer schist, 150
    at Etta Mine, 125
    in Lead area, 86
    at Mount Rushmore, 129, 136
    at Nemo, 20
    at Peerless Mine, 120, 122
    See also Biotite; Muscovite
Microcline, 38-39
    of Bull Moose Pegmatite, 157
    of Etta Pegmatite, 126
    of Harney Peak Granite, 20, 130, 133
    of Iron Mountain Pegmatite, 139
    of Peerless Pegmatite, 122
Milling, gold, 99-100
Mining
    gold, 97-99
    pegmatite, 124, 129
Minnekahta Formation, 9, 15, 28
    at Boulder Park, 61-65
    at Cascade Springs, 164-167
    from Dinosaur Overlook, 117
    at Hot Springs, 161
    in Little Elk Creek Canyon, 112
    at Racetrack, 17, 46, 61
    of Sly Hill, 59
Minnelusa Formation, 15, 27-28, 46, 61, 63, 112
Minnewaste Member, 10
Miocene Epoch, 33, 109
Mississippian age, 9, 77, 79
    at Bear Butte, 54
    at Boulder Park, 61
    at Bridal Veil Falls, 46
    at Jewel Cave, 146
    at Terry Peak, 86
Missouri River, 117

Mitchell, Paul, 120, 122
Monoclines, 112
Monzonites, 30, 80, 82, 86, 96
Morrison Formation, 10, 28, 43, 45, 56, 117
Mount Baldy, 139
Mount Rushmore National Memorial, 7, 20, 117, 124, 129-139
Mount Theodore Roosevelt, 30, 65
Murchisonia, 77
Muscovite, 39
  at Bear Mountain, 20
  of Bull Moose Pegmatite, 155, 157
  of Custer schist, 150
  of Etta Pegmatite, 125
  of Harney Peak Granite, 20
  in Lead area, 92
  at Mount Rushmore, 130, 133, 136
  at Nemo, 20
  of Peerless Pegmatite, 122
  at Strawberry Ridge, 105

Nailhead spar, 149
National Park Service, 144
Needle Eye, 139-144
Needles, The, 7, 15, 20, 22, 139-144
Needles Highway, 139
Nemo, 20, 41, 109-112
Nepheline, 94
Nevada Gulch Road, 83
19th Street, Hot Springs, 161
Niobium, 38, 39, 157
Niobrara Formation, 54
Nonconformities, 7-9, 28, 94
Northern Hills Sanitation Transfer Station, 80

Obsidian, 33
Oligocene gravels, 30
Opeche Formation, 28, 46, 61, 112
Ordovician age, 9, 27, 76, 77
Orman Dam, 85
Orthoclase, 80, 82, 86, 94
Oysters, 45

Pahasapa Formation, 15-17, 27-28, 75, 79
  at Bear Butte, 54
  at Boulder Park, 61
  at Bridal Veil Falls, 46
  caves in, 144, 146-148

from Terry Peak, 85-86
  at White Gate, 112-115
Paleozoic age, 7-9, 11, 15-17, 27, 28, 30, 70, 75
  at Boulder Park, 59-65
  at Cascade Springs, 164
  and Little Elk Creek Canyon, 112, 115
  from Lookout Peak, 46
  under Terry Peak, 86
Paraconformity, 77
Peccaries, 161
Peerless Mine, 120-124
Peerless Pegmatite, 120-124
Pegmatites, granitic, 7, 20, 34-39
  Bull Moose, 153-157
  Etta, 34, 38, 124-129
  Iron Mountain, 139
  at Mount Rushmore, 7, 129, 130, 132, 133
  at The Needles, 7, 144
  Peerless, 120-124
Pennsylvanian age, 9, 61
Pentacrinus, 45
Permian age, 9, 61, 117
Phanerozoic interval, 24-34
Phonolite, 30, 46
Phyllites, 7, 24, 86, 88
Piedmont, South Dakota, 112
Pierre Shale, 54, 120
Plagioclase, 23, 80, 122, 133, 139. *See also* Albite
Pleistocene age, 10, 30, 34
  at Buffalo Gap, 158
  at hot springs, 161, 164, 169
Pliocene age, 30
Ponderosa Motel, 92
Poorman Formation, 86, 92-94, 96
Portland Cement, 117
Positive regions, 11
Powder River Basin, 11, 27
Precambrian age, 7, 11, 15, 18-24, 30, 72, 75
  of Harney Peak, 20, 85, 117
  in Lead area, 86-94
  at Nemo, 20, 109
  at Peerless Mine, 122
  at Strawberry Ridge, 101, 104, 105, 109
  from Terry Peak, 86
  at Tomahawk Lake Country Club, 106
Pringle, 150, 155
Pyrite, 82, 97
Pyroxene, 23, 30, 83, 86, 94
Pyrrhotite, 97

Quartz, 20, 23, 30, 34, 38, 39, 72-75
  at Bear Mountain, 20
  of Bull Moose Pegmatite, 155, 157
  of Custer schist, 149-153
  of Cutting Stock, 80
  at Etta Mine, 125, 126, 127-129
  at Iron Mountain, 139
  in Lead area, 88, 92, 97
  at Mount Rushmore, 130, 133, 136
  at Nemo, 20, 109
  of Peerless Pegmatite, 122
  of Sheep Mountain Stock, 65
  of Terry Peak, 86
Quartzite, 15, 24
  of Bull Moose Pegmatite, 155
  in Lead area, 88, 97
  at Nemo, 109
  on Strawberry Ridge, 105
Quincy Avenue, Rapid City, 117

Racetrack. *See* Red Valley
Rainbow Arch, 63
Rainfall, 13, 34, 116, 117
Rapid Canyon, 117
Rapid City, South Dakota, 41, 104, 112, 117
Rapid Creek, 117
Recent era, 18, 24
Receptaculites, 77
Red Gate, 112, 115
Red Valley, 9, 17-18, 61
  from Dinosaur Overlook, 117
  and Little Elk Creek Canyon, 116
  from Lookout Peak, 43, 45-46
  from Sly Hill, 56-59
  from Terry Peak, 85
Rhyolite, 30, 51-54, 65, 88, 96
Rocky Mountains, 11, 29, 75
Ross shaft, 97
Roubaix, 33
Roy Street, Keystone, 120
Rushmore Pegmatite, 120-124
Rushmore Tramway, 125

Sandstones, 9, 22, 24, 136
  in Deadwood Formation, 27, 70, 72-75, 94, 96, 101, 104, 115
  of Ellison Formation, 88
  of Fall River Formation, 10, 18, 55, 56

of Lakota Formation, 10, 18, 55, 56, 117
of Laramide Orogeny, 28-29
of Minnelusa Formation, 61
of Sundance Formation, 10, 18, 28, 43, 55
Schists, 23-24, 30
of Bull Moose Pegmatite, 155
at Custer, 7, 149-153
at Etta Mine, 125-126
of Iron Mountain, 139
in Lead area, 7, 86, 92, 94, 97
at Mount Rushmore, 129-130, 132-133, 136
at The Needles, 144
at Nemo, 109
at Peerless Mine, 122-124
of Strawberry Ridge, 101, 104-105
Sculptures, at Mount Rushmore, 130-132, 133, 136-139
Shales, 9, 18, 22, 23, 24, 136
of Deadwood Formation, 70, 94, 101, 115
of Englewood Formation, 27, 77
of Fall River Formation, 10
of Graneros Formation, 59
of Gypsum Spring Formation, 9-10
of Lakota Formation, 10, 56
of Laramide Orogeny, 28-29
at Minnelusa Formation, 61
of Morrison Formation, 10
of Opeche Formation, 61
Pierre, 54, 120
of Spearfish Formation, 9, 17, 43, 157-158
of Sundance Formation, 10, 28, 43, 45, 157, 167
of Unkpapa Formation, 10, 157-158
of Winnipeg Formation, 76
Sheep Mountain Stock, 65
Sideroplesite, 92, 97
Silicates, 9, 24, 28, 82, 133, 149
Sillimanite, 23, 136, 150, 153, 155
Siltstone, 9, 45, 56, 61
Silurian period, 9, 27, 77
Silver, 39, 82, 97, 100
Slate, 24, 109
Sly Hill, 56-59
Snails, 27, 161
Spearfish Canyon, 46-51, 85
Spearfish Canyon Road, 46

Spearfish Creek, 46-49
Spearfish Formation, 28
at Boulder Park, 61
at Buffalo Gap, 157, 158
at Cascade Springs, 164-167
at Hot Springs, 161-164
of Red Valley, 9, 17, 43, 46, 56, 117
Spearfish, South Dakota, 43-46
Spodumene, 39, 125, 126, 127-129
State Cement Plant, 41, 104, 112
Staurolite, 23, 122, 125, 153
Stocks, 30, 37, 65, 80-83
Stone Mountain, Georgia, 136
Stop 1, 43-46
Stop 2, 46-51
Stop 3, 51-56
Stop 4, 56-59
Stop 5, 59-65
Stop 6, 65
Stop 7, 65-70
Stop 8, 70-75
Stop 9, 75-80
Stop 10, 80-83
Stop 11, 83-86
Stop 12, 86-94, 104
Stop 13, 94-100
Stop 14, 100-104
Stop 15, 104-105
Stop 16, 105-109
Stop 17, 109-112
Stop 18, 112-117
Stop 19, 117-120
Stop 20, 120-124
Stop 21, 124-129
Stop 22, 129-139
Stop 23, 139
Stop 24, 139-144
Stop 25, 144-149
Stop 26, 149-153
Stop 27, 153-157
Stop 28, 157-161
Stop 29, 161-164
Stop 30, 158, 164-169
Strawberry Hill Campground, 100
Strawberry Hill Mining Company, 65
Strawberry Ridge, 41, 100-105, 109
Sturgis, South Dakota, 56-59, 63, 117
Sundance Formation, 10, 17, 18, 28
of Bear Butte Circus, 55
at Buffalo Gap, 157

from Dinosaur Overlook, 117
from Sly Hill, 43-45
Sundance Mountain, 18
Sundance Peak, 11
Sundance, Wyoming, 27
Sylvan Lake, 20, 139
Syringopora, 79, 112

Taconite, 41, 109
Tantalite-columbite, 120, 157
Tantalum, 38, 39, 157
Temperature
climatic, 13
of hot springs, 168
Terry, Alfred H., 83-85
Terry Peak, 15, 30, 83-86
Terry Peak Chairlift, 83
Tin, 38, 39, 125
Tinton, 30, 34
Tomahawk Lake Country Club, 33, 105-109
Tourmaline, 39
of Bull Moose Pegmatite, 157
of Custer schist, 150
of Etta Pegmatite, 125
of Harney Peak Granite, 20
of Iron Mountain Pegmatite, 139
of Peerless Pegmatite, 122
Travertine, 149
Triassic age, 9, 56, 61, 157, 161
Trilobites, 27, 46, 72, 115

University Avenue, Hot Springs, 161
Unkpapa Formation, 10, 117, 158
U.S. Government, 100

Visitors' Center, Mount Rushmore, 129, 130
Volcania breccia, 106-109
Volcanic glass, 33, 106
Vugs, 17, 79

Washington (George) sculpture, 130, 133
Weathering, 15
West Boulevard, Rapid City, 117
Whirlybird Rides, 125
White Gate, 112-117
White River Formation, 120
Whitewood Creek, 65, 70, 75

Whitewood Formation, 27, 75,
  76-77, 115
Whitewood, South Dakota, 18,
  85
Williston Basin, 11, 27
Winnipeg Formation, 75-76,
  115
Wolframite, 153
Wonderland Cave, 116

Yates shaft, 97-99